FOLLOWING OIL

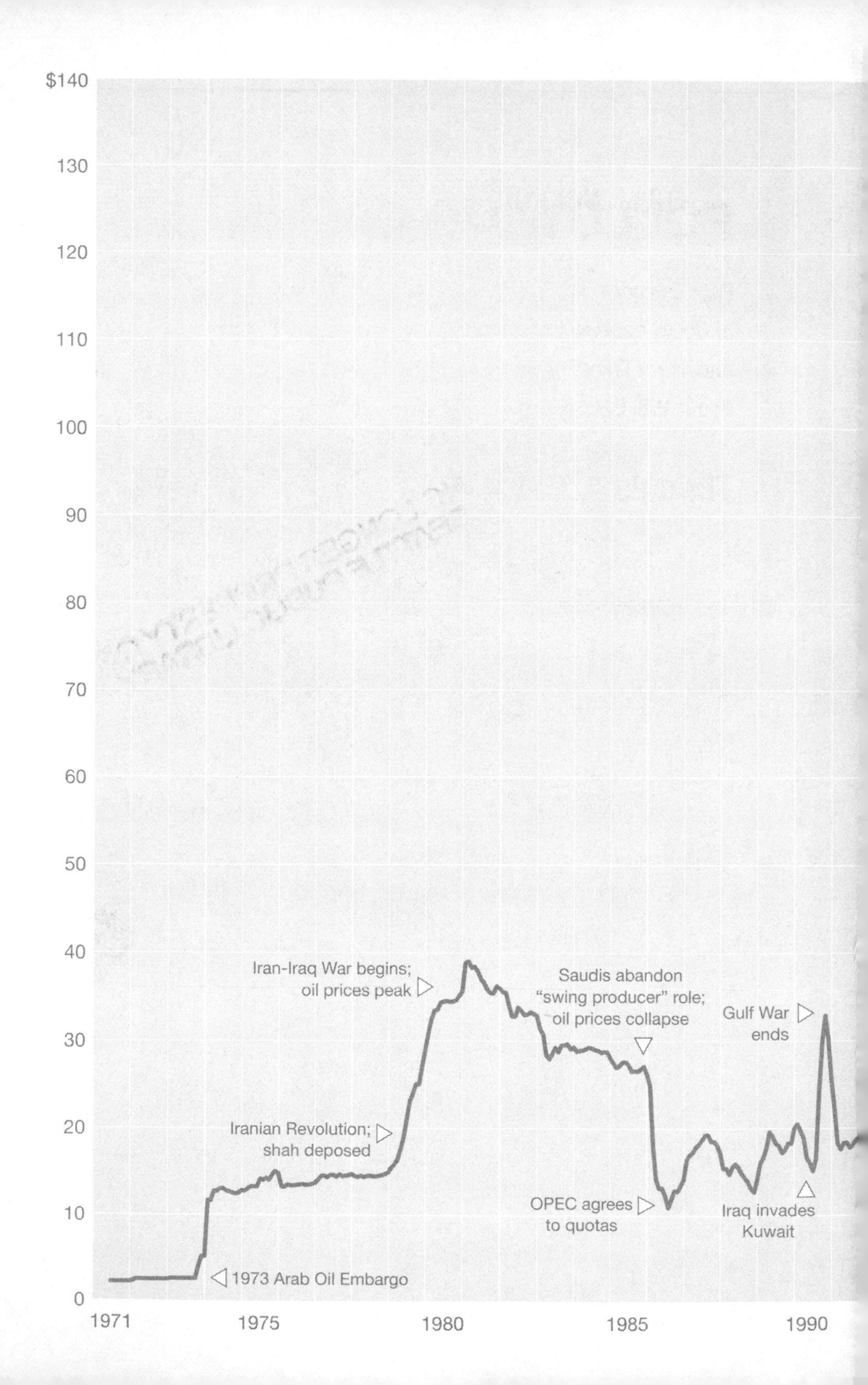

$140
130
120
110
100
90
80
70
60
50
40
30
20
10
0
1971
1975
1980
1985
1990
1973 Arab Oil Embargo
Iranian Revolution; shah deposed
Iran-Iraq War begins; oil prices peak
Saudis abandon "swing producer" role; oil prices collapse
OPEC agrees to quotas
Gulf War ends
Iraq invades Kuwait

FOLLOWING OIL

Four Decades of Cycle-Testing Experiences and What They Foretell about U.S. Energy Independence

Thomas A. Petrie

UNIVERSITY OF OKLAHOMA PRESS : NORMAN

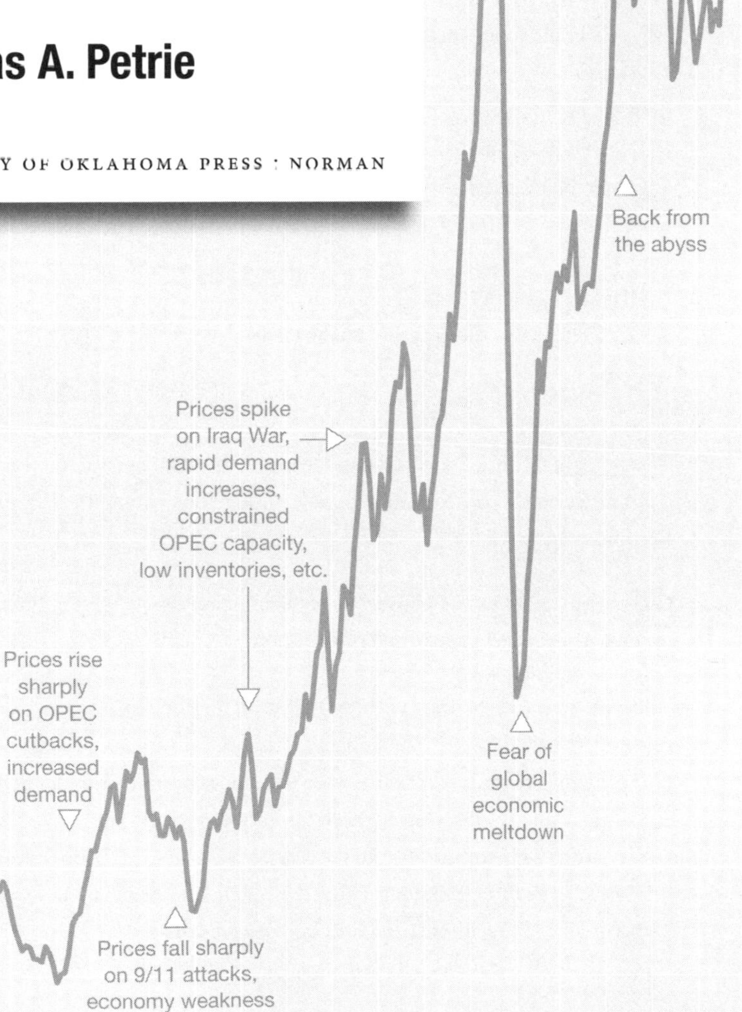

Library of Congress Cataloging-in-Publication Data

Petrie, Thomas A.

Following oil : four decades of cycle-testing experiences and what they foretell about U.S. energy independence / Thomas A. Petrie.

pages cm

Includes bibliographical references and index.

ISBN 978-0-8061-4420-7 (hardback : alk. paper)

1. Petroleum industry and trade—United States—Finance. 2. Petroleum reserves—United States. 3. Energy development—United States. 4. Renewable energy sources—United States. I. Title.

HD9565.P3497 2014

338.2'72820973—dc23 2013034848

The paper in this book meets the guidelines for permanence and durability of the Committee on Production Guidelines for Book Longevity of the Council on Library Resources, Inc. ∞

2 3 4 5 6 7 8 9 10

Dedicated to the memory of

FREDERICK R. MAYER

and

CORTLANDT "CORT" S. DIETLER

Two pioneering energy sector entrepreneurs

who led by example

CONTENTS

ILLUSTRATIONS

Figures

Maps

PREFACE

The news of America's revolution in unconventional oil and gas supplies is ubiquitous. One cannot read the *Wall Street Journal*, the *New York Times*, the *Economist*, or a host of other press sources and periodicals without encountering frequent accounts of how this dramatic change in supply prospects is unfolding. There is renewed broad optimism about the expected favorable impacts on U.S. energy security, trade balances, and job creation. In November 2012, the International Energy Agency (IEA) joined the chorus by releasing its updated *World Energy Outlook 2012*. This organization is based in Paris and was created in the aftermath of the first oil crisis of the early 1970s by the developed consuming nations to centralize the study and assessment of global trends. Its latest report represents a major acknowledgment of the upturn in U.S. oil and gas production stemming from the application of new technologies to exploit fossil fuel resources previously considered unreachable or uneconomic. As the IEA report describes, these developments are redrawing the global energy maps in a variety of ways.

In 2010, China displaced the United States to become the world's largest consumer of energy from all sources. Despite a slower-growing economy, by 2015 China is expected to top the United States again, this time as it becomes the world's largest importer of oil. In the next few years, U.S. oil imports are expected to decline to well under half the levels prevailing in the past decade. The increase in domestic production along with reduced consumption as a result of improved automobile mileage standards and other conservation measures could be sufficient to almost totally eliminate America's need for oil imports. In fact, it could even create the need for congressional action to reconsider legalizing non–North American oil exports. By act of Congress, the export of crude oil has been essentially off-limits since the Arab oil embargo of the early 1970s. This does not apply to refined products. However, optimal economics will likely require the flexibility to export some crude oils more suitable for foreign refiners and the import of other lower-value crudes for which the U.S. refineries are best designed to run. The revolution in development of unconventional gas is also providing benefits in terms of reduced emissions of carbon dioxide. As the *Wall Street Journal* recently reported, over the past five years, the United States has made more progress than any of the countries that are signatories to the Kyoto Accord.

The idea of writing a book about my experiences in the petroleum industry began to take shape sometime around the middle of the first decade of what is now a four-decade career, first as an oil and gas investment analyst and subsequently as an investment banker and strategic advisor on petroleum sector mergers, acquisitions, and financings. By that time, the historic events surrounding the rise to economic power of the Organization of Petroleum Exporting Countries (OPEC) had already underscored the central role of energy in the national security calculus of the United States and its Western allies. At almost the same time, the emerging confrontations over development of Alaskan North Slope oil and the pioneering efforts to pursue large North Sea oil projects convinced me that these were highly suitable case studies for assessing the effectiveness of capitalism in energy development on a national, and even global, scale.

As a result, during the course of my work as an energy securities investment analyst, I began to keep copies of my oil research reports, key related public documents of the times, personal calendars detailing my travels, and other relevant files. In the late 1980s, I intended to title the book *Inflection Points*, as I had by then observed firsthand enough of the petroleum sector's chronic cyclicality, as illustrated on the annotated oil price chart on pages xvi–xvii, to realize that such turning points did much to define the scope and character of each subsequent unfolding era. Time and again I was observing how downturns, setbacks, and reversals set up new opportunities for thoughtful capitalists and prudent risk takers.

More recently, I evolved to terming it *Following Oil* because this book is about two notions. First, it addresses what I believe I have learned from following the petroleum sector as a securities investment analyst dealing with institutional investors for almost two decades, as well as another two decades as an energy investment banker. Second, it speaks to the examination of what energy sources could eventually become available on a relevant scale following an inevitable shrinkage in oil availability, about which there has been much recent political debate centering on the need for growing contributions from renewable, sustainable, and green energy sources.

The format is to review roughly chronologically that learning process in which I have been so fortunate to be immersed. To do this, I have sought to relate a series of stories about the actions and contributions of key people and events that I have personally witnessed. They are by no means a comprehensive accounting of all I have experienced. Rather, they represent a selection of some of the more noteworthy recollections about major events or situations that demonstrates the essence of unfolding trends for energy. Finally, these vignettes often reveal time-tested insights regarding the major drivers of petroleum activities and economics. The resulting focus is on how energy markets function when impacted by domestic and global trends in economic growth, evolving demographics, a maturing resource base, variable national energy policies, and dynamic changes in technology and geopolitical forces.

ACKNOWLEDGMENTS

I am indebted to many individuals whom I wish to acknowledge for their input and support over the several years of my work on this project. With respect to the combined efforts it took to fashion my initial manuscript into a published book, I especially thank Byron Price and Chuck Rankin of the University of Oklahoma Press. Early on and throughout the process, they provided encouragement and advice that was invaluable in shaping my efforts. Steven Baker's oversight of the editing is deeply appreciated, as are Sally Bennett Boyington's soundly reasoned editorial improvements in the manuscript. Thanks as well to Carol Zuber-Mallison for her talented translation of my notional figures and maps into informative images. I appreciate the accommodation of my Petrie Partners colleagues Jon Hughes (also a reader), Mike Bock, Andy Rapp, Scott Baxter, Brad Marvin, and Jacob Nagy.

None of this would have been possible without the consistent support, attention to detail, and enthusiasm of Janeen Hogan, my executive assistant of almost thirty years. Her ability to catch my misses, anticipate obstacles, and resolve issues was critical to this endeavor.

I extend my gratitude to Petrie Partners analysts Jesse Irvin, Pete Williams, and David Knop for their availability to run down and check key facts. Chris Peck provided research that brought "back to life" the details of the regulatory environment and changing geopolitics of the 1970s, as well as the M&A consolidation and commodity fluctuations of the 1980s. Barry McKennitt, who deserves credit for almost single-handedly ensuring the continued functioning of the National Association of Petroleum Investment Analysts, was helpful in reconstructing various details of that organization's meetings and other activities in the 1970s and 1980s. Linda Lebsack was especially helpful in locating key out-of-print copies of petroleum sector histories. My former First Boston colleague, Art Smith, brought clarity and focus to the data depicting the petroleum industry's waves of M&A consolidation, and provided useful details of that era. Steven Andrews was particularly helpful on issues involving global peak oil concerns.

I wish to thank Sara Anton and her family for permission to include her late husband, Albert J. Anton, Jr.'s, account of the oil analysts' travails escaping the North Tower of the World Trade Center on September 11, 2001.

I am especially grateful for the constructive feedback (negative and otherwise) of those who read the manuscript in its various stages of evolution. These include Jurek Antoszewski, Jim Clark, Jim Galbreath, Harold Hamm, Harold Korell, Hal Logan, Bill Mayer, Nikos Monoyios, Tony Pace, Jay Precourt, Trevor Rees-Jones, Frank Reinhardt, Arlie Sherman, Dick Strong, Byron Wien, Dan Yergin, and Brian Young.

I am indebted to a number of industry colleagues, mentors, and friends who over these many years have contributed critical insights about the economics, geopolitics, corporate and board dynamics, and interpersonal relations that so often are the drivers of energy sector events and outcomes. In part, these include Phil Anschutz, Bob Boswell, Pat Broe, Ray Brownlie, Steve Chazen, Robert Day, Craig Drill, the Honorable Don Evans, David Leuschen, Martin Lovegrove, Ken Peak (now deceased), Raymond Plank, Bill Randol, Dr. Bill Scoggins, Jim Volker, Jim Wallace, Don Wolf, Marvin Wolf, George Wood, and Paul Zecchi. The inputs of everyone cited, as well as countless others who have in various endeavors impacted my career, are highly valued. I alone, however, am responsible for any shortfalls or inaccuracies.

Finally, the patience, love, and guidance of my wife, Jane, has sustained me throughout this project and inspired me to distill insights from a lifetime's adventure.

ABBREVIATIONS

ANWR	Arctic National Wildlife Refuge
bbl	barrel
Bbpd	billion barrels per day
BOEM	Bureau of Ocean Energy Management
bpd	barrels per day
BTU	British thermal unit
CBO	Congressional Budget Office
CEO	chief executive officer
CFO	chief financial officer
CFTC	Commodities Futures Trading Commission
DOE	Department of Energy
E&P	exploration and production
EIA	Energy Information Administration
EOR	enhanced oil recovery
EPA	Environmental Protection Agency
EPCA	Energy Policy and Conservation Act
FEA	Federal Energy Administration
FPC	Federal Power Commission
FTC	Federal Trade Commission
GHG	greenhouse gas
IEA	International Energy Agency
IPAA	Independent Petroleum Association of America
IPO	initial public offering
JV	joint venture
LEED	Leadership in Energy and Environmental Design
LNG	liquefied natural gas
LTCM	Long-Term Capital Management
M&A	mergers and acquisitions
MEOW	moral equivalent of war

MLP	master limited partnership
MMbpd	million barrels per day
MTBE	methyl tertiary butyl ether
NAPE	North American Prospect Expo
NAPIA	National Association of Petroleum Investment Analysts
NATO	North Atlantic Treaty Organization
NGL	natural gas liquids
NGPA	Natural Gas Policy Act
NIMBY	"not in my backyard"
NO_x	nitrogen oxides
NRC	Nuclear Regulatory Commission
NREL	National Renewable Energy Laboratory
OECD	Organisation for Economic Co-operation and Development
OPEC	Organization of the Petroleum Exporting Countries
S&P	Standard & Poor's
SEC	Securities and Exchange Commission
SERI	Solar Energy Research Institute
SO_x	sulfur oxides
SPE	Society of Petroleum Engineers
SWRP	Subsea Well Response Project
USGS	United States Geological Survey
USMA	United States Military Academy
VLCCs	very large crude carriers
WIP	whip inflation now

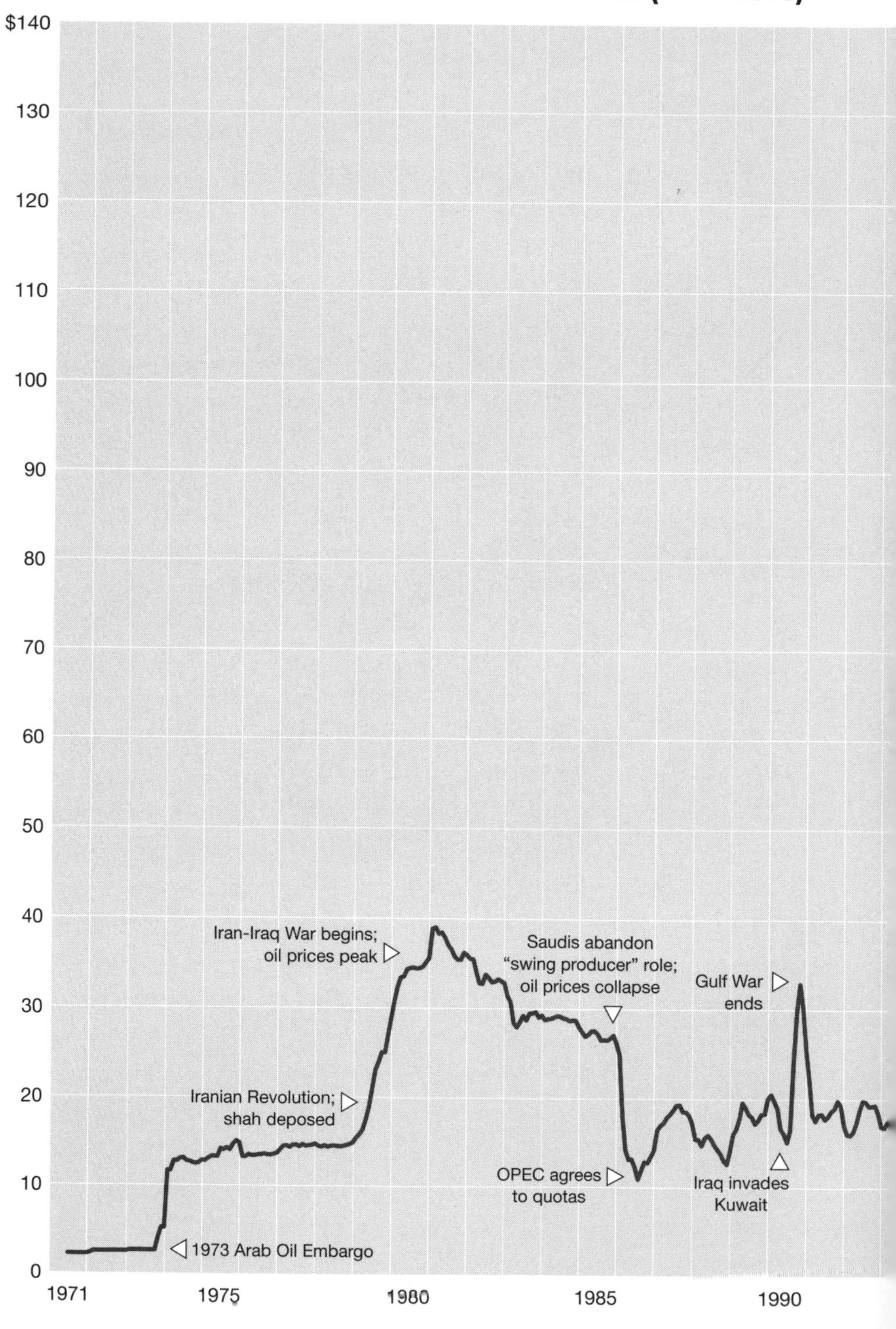

WORLD EVENTS AND OIL PRICES–NOMINAL PRICING (1971–2013)
$140
130
120
110
100
90
80
70
60
50
40
30
20
10
0
1971
1975
1980
1985
1990
Iran-Iraq War begins; oil prices peak
Saudis abandon "swing producer" role; oil prices collapse
Gulf War ends
Iranian Revolution; shah deposed
OPEC agrees to quotas
Iraq invades Kuwait
1973 Arab Oil Embargo

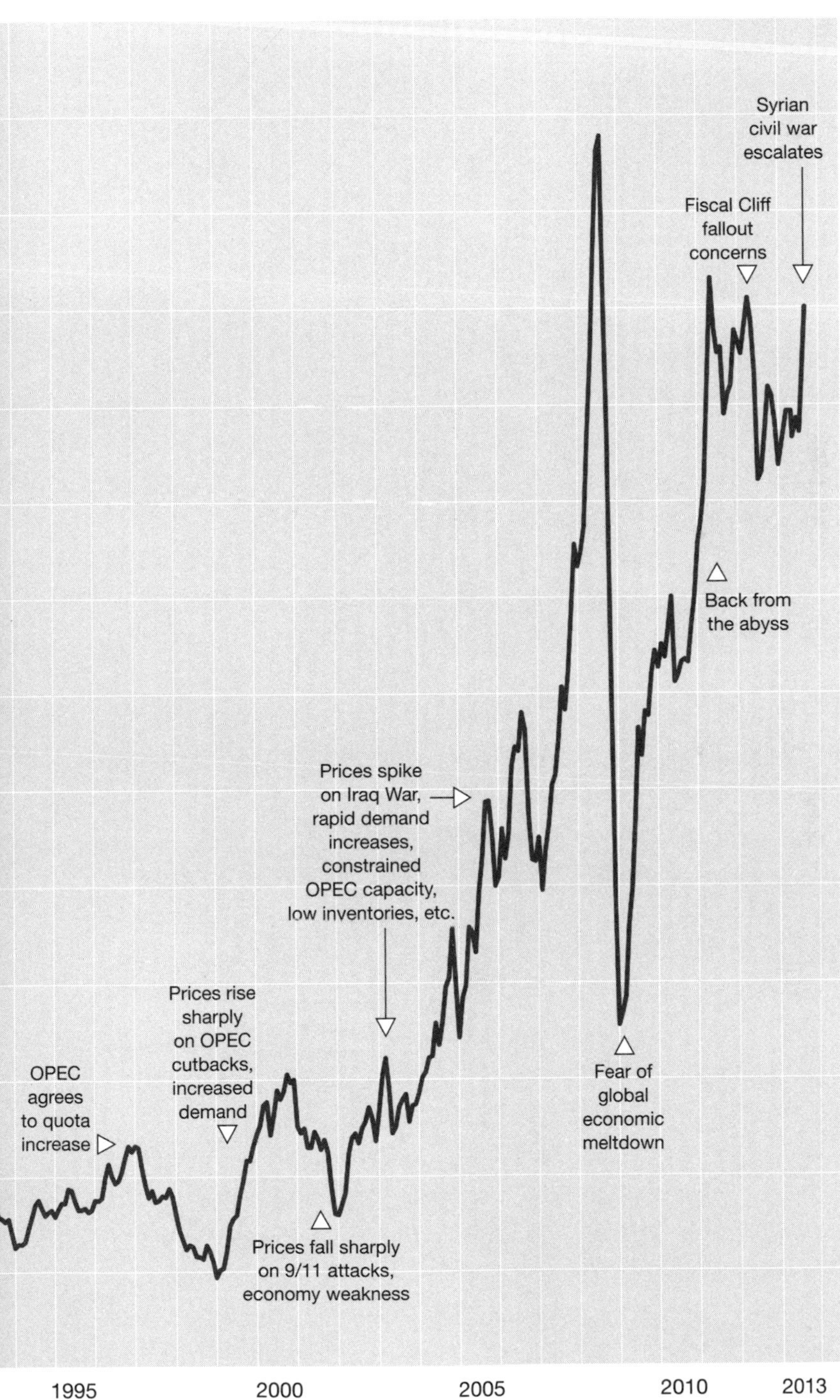

Syrian civil war escalates
Fiscal Cliff fallout concerns
Back from the abyss
Prices spike on Iraq War, rapid demand increases, constrained OPEC capacity, low inventories, etc.
Fear of global economic meltdown
Prices rise sharply on OPEC cutbacks, increased demand
OPEC agrees to quota increase
Prices fall sharply on 9/11 attacks, economy weakness
1995
2000
2005
2010
2013

PART I

CYCLE-TESTING EXPERIENCES, 1967–2013

ISRAEL'S SURPRISE ATTACK DURING THE SIX-DAY WAR, JUNE 5-10, 1967

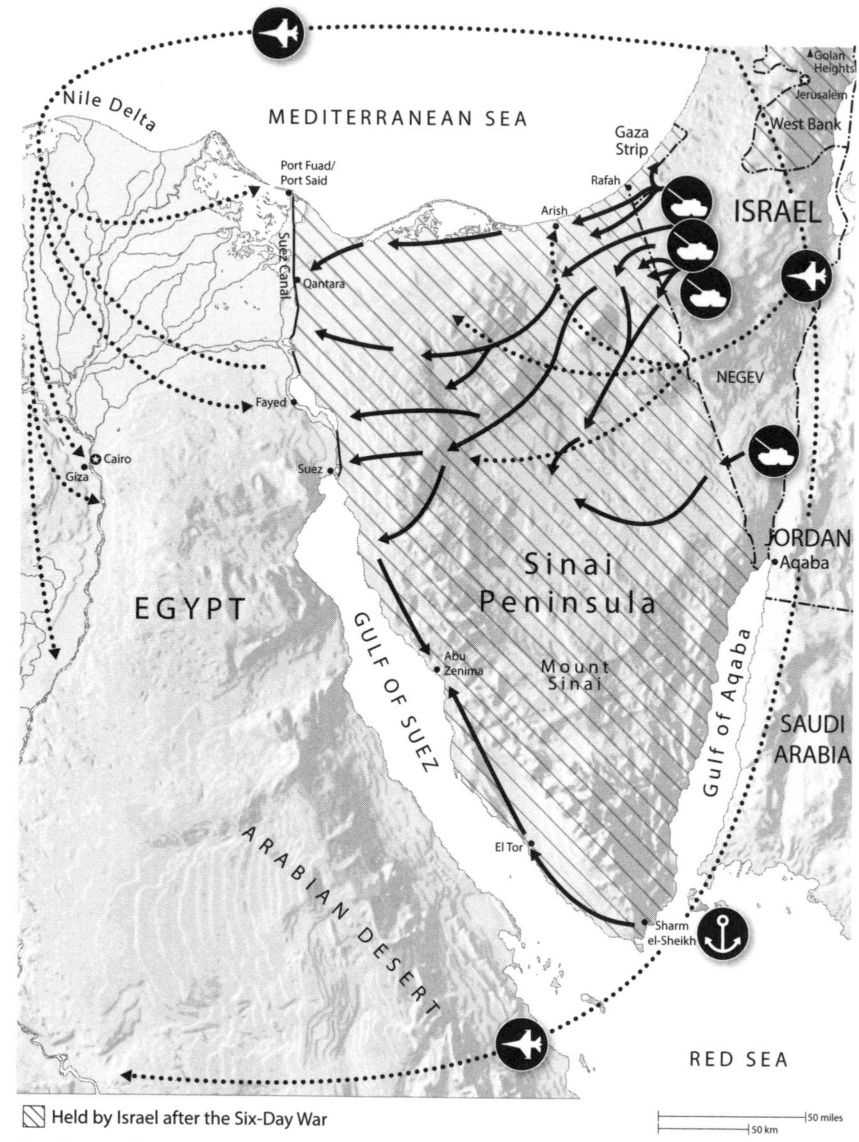

Held by Israel after the Six-Day War

Israel captured:
- Gaza Strip and the Sinai Peninsula from Egypt
- West Bank (including East Jerusalem) from Jordan
- Golan Heights (not shown) from Syria

Chapter 1

EARLY ENCOUNTERS WITH THE OIL SECTOR

The Six-Day War, June 1967

It could have been the first oil crisis
but wasn't.

As the Corps of Cadets formed up for the final Graduation Parade by myself and 582 other members of the U.S. Military Academy Class of 1967 on June 6, loudspeakers in the barracks areas announced that starting the prior day, Egypt and its other Arab allies were once again at war with Israel. While I did not fully appreciate the ramifications of what I was then observing, the fallout of this conflagration would be a factor that set the stage for a variety of global energy issues that I would later encounter over an ensuing four crisis-punctuated decades of my career as a petroleum analyst and energy investment banker.

Middle East tensions had been building for weeks, so the outbreak of hostilities was not entirely a surprise to the intelligence services of virtually all the major affected countries. While the Arab armies were still marshaling, Israel mobilized quickly and established momentum that enabled it to neutralize the opposing forces. A key element of its strategy was a decisive strike by an air-force-supported armored column driving across the Sinai Desert, with the goal of closing the Suez Canal. Israel achieved this objective in literally a matter of a few days, thereby inflicting a significant penalty on Egypt for its leadership of the Arab coalition. This included entrapment of one-third of the country's army in the Sinai Desert, loss of canal transshipment fees, occupation of the Sinai, disruption of tourism, and other indirect economic effects. On the sixth and final day of the conflict, other Israeli forces pushed up onto the Golan Heights and thus extended Israel's perimeter to the northeast with Syria to much more defensible terrain.

This redrawing of the "Near Eastern" Middle East map did much to set the stage for an ongoing period of friction and agony that has characterized many, though not all, of the confrontations occurring in the region in the ensuing decades.

By many measures, what we now know as the "Six-Day War" of 1967 could also have given rise to the first global oil crisis. The short-haul route for Middle East oil going to Europe and North America was cut off, as would remain the case for over a decade. However, an actual crisis in terms of oil supply, it was not to be. Surplus U.S. oil production capacity readily available from a series of large fields discovered in the 1920s and 1930s (mainly in Texas and Louisiana) along with increased oil output by Iran and Venezuela were ample to meet both U.S. and European needs. On the first news, oil prices initially increased by about 10 percent but held there only briefly. They then rapidly retreated as the valves of non-Arab foreign and U.S. oil wells in West Texas and Louisiana were opened to ramp up output. These increases in output were interim measures that provided flexibility until the Arab embargo subsided and new supertankers, known as very large crude carriers (VLCCs), could be built to capitalize on the economies of scale needed to haul Middle East oil all the way around Africa to supply Atlantic basin markets.

Consequently, the return to the "old normal" of U.S. oil prices below three dollars per barrel (and OPEC oil below two dollars per barrel) was swift. For the moment, the message to global oil markets was that it would take more than a Middle East military conflagration to upset the economic order of the petroleum sector that had evolved post–World War II. Given a sixfold increase in tanker rates caused by the initial shrinkage in effective shipping capacity owing to the much-lengthened journey around Africa to reach Western markets, the build-out and launching of these highly economical ships occurred relatively quickly over the next three-plus years. This was a remarkable demonstration of how relatively unregulated markets can and often do respond to strong price signals.

However, some six years later, the balance of global petroleum power would indeed change dramatically. Spare U.S. production capacity was by then exhausted and had entered what soon proved to be a long-term irreversible natural decline from peak domestic production starting in 1970.

The next Arab/Israeli war—at the time of Yom Kippur in October 1973—set in motion forces that resulted in a sustained increase in oil prices to more than triple their previous level. Price volatility would continue periodically to whipsaw global markets for years (and ultimately decades) thereafter. For consumers, U.S. oil priced at three dollars per barrel, and with it, cheap gasoline prices, became only a fond memory. Furthermore, OPEC, long considered a "toothless tiger" since its formation in 1961, was no longer so characterized.

My Grounding in the Petroleum Sector

Early lessons learned.

In the late spring of 1971 in the middle of this historic transition, I entered the profession of petroleum investment analysis. Following my commissioning as a second lieutenant, I completed the U.S. Army's Ranger School at Fort Benning, Georgia, as well as a couple of other officer training courses and then proceeded to two overseas tours, first in Germany and then in Vietnam. After returning to the United States for one more tour of duty, which concluded my contractual commitment to the U.S. Army, I decided not to pursue a career in the army. Having earned a masters degree in business administration through Boston University's overseas program in Frankfurt, Germany, I was interested in pursuing a career in the investment sector. While jobs were still scarce in 1971 because of the lingering effects of the 1970 recession, I was fortunate to land an entry-level securities investment analyst position with Colonial Management Associates in Boston.

Colonial was an investment advisory firm whose clients included the endowment funds of Massachusetts Institute of Technology, Dartmouth College, and Massachusetts General Hospital, as well as a family of mutual funds. The firm was formed by James H. Orr, Sr., in the early 1930s. In the late 1920s, Mr. Orr had been a portfolio manager with the investment management subsidiary of the engineering and construction firm Stone & Webster. In the aftermath of the 1929 stock market crash, he had been given the task of finding a buyer for that operation. After many months of striking out, Orr reported to his superiors that given the depressed securities markets there were no other buyers, but he personally was prepared to assume responsibility for the investment portfolios. By 1932, he was in business for himself, and with shrewd trading and investment decisions, by mid-decade he was on his way to running a viable portfolio management enterprise. Sometime around 1934, he bought Tampa Electric bonds for less than the arrears interest and not too long thereafter was rewarded with their redemption at par with the full interest paid. In the deal with Stone & Webster, Mr. Orr was also aided by his having obtained control of the Rail and Light Securities Fund. Organized in 1895 along the lines of a Scottish closed-end investment trust, this fund gave him an arguable claim to be the first mutual fund (or certainly one of the first) in America. With that tagline and the fund's reconstitution as the Colonial Fund, Mr. Orr had entered the investment business at close to the Depression's market bottom.

In 1936, his reputation, much enhanced by a series of successful judgments (such as buying the Tampa Electric bonds) in a tough investment environment, resulted in an invitation for Orr to become the first public director of First Boston Corporation. This was the investment banking arm then being spun off from the First National Bank of Boston, pursuant to the Glass-Steagall Act. During World War II, George Woods, a senior First Boston executive, took a leave of absence to join the War Production Board, and it was Orr's task to take the train from New York to Washington following each board meeting to brief Woods on First Boston matters. In the postwar years, he built Colonial into a respected player among Boston money managers, and when Jerry Tsai sold his Manhattan Fund in 1968 for a stellar price, Orr and his partners followed shortly thereafter with a sale of Colonial Management at fifty-five times earnings.

All this occurred before I joined the firm, but hearing the story and knowing the principals involved opened my eyes to the exciting possibilities if one could combine effective judgment, a focused strategy, and a keen sense of timing in the financial sector. Jim Orr was my first mentor in the investment business. His advice and counsel over three decades was invaluably inspiring and remains deeply appreciated.

What I realize today is how fortuitous was the timing of my joining the firm. When I started as an inexperienced analyst, responsible for developing an understanding of publicly traded securities in the energy sector, this category of investments had been languishing for two years and was not showing signs of an imminent upturn. The stock market's aversion to oil sector investments was due in part to lingering effects of the prior year's recession. It also reflected rising investor concerns about the security of oil company ownership in foreign exploration and production (E&P) concessions. This was all in the aftermath of Colonel Moammar Khadafy's bloodless coup that deposed Libya's King Idris in 1969. While still serving in the U.S. Army in Germany, I had become well aware of the Khadafy takeover because it brought to an end the low-cost vacations by U.S. Army personnel looking to enjoy the North African side of the Mediterranean coastline. More importantly, Khadafy's ascension to power proved to be the beginning of a series of adverse changes in the economic terms for oil companies operating throughout the oil-exporting countries. In succession, Iraq, Venezuela, Nigeria, and Kuwait all weighed in to exercise their sovereignty. Combined with the early evidence of U.S. production shortfalls in 1971–72, these changes set the stage for the first oil crisis in 1973.

In retrospect, I was given the opportunity of a lifetime. John McNiece, my manager at Colonial, encouraged me to take the time available to learn all I could about the energy industry generally and, most particularly, about the petroleum sector. Over the following year, I was able to travel widely throughout North America to meet many members of senior management and become familiar with different energy companies. This experience allowed me to build a framework for understanding many of the corporate strategies and business models of the energy companies as well as the technical and fundamental drivers of performance and investment attractiveness of petroleum enterprises.

The first trip was to Marathon Oil in Findlay, Ohio. There I met several members of the company's management, including the chief financial officer, Elmer Graham. A full decade later as a corporate defense advisor, I was to spend intense time with Graham when Mobil Oil launched a hostile tender offer for the company. At this time, however, Marathon and its partner, Phillips Petroleum, were in the startup phase of a project to ship liquefied natural gas (LNG) from Alaska's Kenai Peninsula to Japan. While in percentage terms this was not a major asset of the company, in the then-prevailing environment of stable oil and gas prices as well as flat conventional oil and gas production volumes, a new and growing LNG export project was one of the more noteworthy drivers of the company's earnings outlook for the next several years. In addition, it afforded me the chance to gain an appreciation of the economics of LNG in the Pacific Rim, something that has been very helpful in assessing transactions involving other companies in subsequent decades. I also started to learn about the Yates field, a true West Texas giant oil accumulation. Having a deep understanding of this "crown jewel" asset of Marathon's would provide a major career-advancing opportunity for me a decade later.

My next visit that summer was to Louisiana Land and Exploration Company (LL&E), where I began to appreciate the value of owning undeveloped petroleum mineral rights in a strategic location when the necessary technology emerges to identify and exploit attractive hydrocarbon resources. LL&E was formed in the late 1920s and benefited from the advent of geomagnetic surveys and subsequently from seismic technology to locate previously unrecognizable subsurface traps of oil and gas in the marshlands of southern Louisiana. From 1930 to well into the 1950s, the company was able to benefit from the leasing of its extensive land holdings to Texaco and other major oil companies that brought both the financial capital and the exploration expertise to develop significant production on LL&E leases. As a royalty holder with the operating and capital costs that were allocated to the working-interest operators, LL&E enjoyed impressive returns and free cash flow.

Now, however, the task at hand was to develop a research perspective on LL&E's new strategy to transform itself into a more traditionally structured operating oil company that would reinvest its royalty cash flows in drilling new wells as a working-interest owner. Toward that end, the company in 1971 was benefiting from a noteworthy success involving its participation with Standard Oil of New Jersey (now Exxon) in the discovery and development of the Jay field in what was known as the Smackover Trend in Florida. This field caught my attention, partly because the Jay field was the largest onshore U.S. discovery of oil in over two decades (since the 1948 opening of the Sacroc field in West Texas). It also piqued my interest because it was located in Florida's panhandle, not far from where I had endured and survived the "swamp" phase of Army Ranger training in 1967. I had hiked many hundred miles in combat gear over that kind of ground while alert to avoid poisonous snakes and alligators. However, I have since come to appreciate that worthwhile economic treasure could lie beneath otherwise very unattractive and challenging parts of the earth's surface. This is a lesson that was reinforced repeatedly over subsequent years on trips to regions ranging from the Alaskan and Canadian Arctic to Colombia and Peru in South America. These trips included numerous excursions to other remote regions in both the Eastern and the Western Hemispheres of the globe. Much later, in 1983, my understanding of the fundamental positioning of LL&E paid dividends. Then I was part of the First Boston mergers and acquisitions (M&A) defense team that helped defeat an unsolicited hostile bid for the company. It involved turning back an opportunistic proxy contest initiated by Delo Caspary and two of the Hunt brothers (Nelson Bunker and William Herbert).

In the early fall of 1971, I attended an oil analysts meeting at Club 21 on Fifty-Second Street just west of Fifth Avenue in New York City called by Leon Hess, CEO of Amerada Hess Corporation. This was my introduction to one of the more interesting, challenging, and "out of the box"-thinking oil executives I have met during my career. The Leon Hess story is the stuff of legend, and it did much to convince me that I had found a profession in which one could be passionate about developing and pursuing insightful investment conclusions.

In the 1930s, Leon Hess drove a fuel truck for the distribution business that his father, Mores Hess, had developed in New Jersey. In the Great Depression year of 1933, the business went bankrupt, and Leon, still in his twenties, reorganized it and put it on a growth path by building a terminal in Perth Amboy, New Jersey. During World War II, he served in Europe as a fuel officer in General George Patton's Third Armored Division. There he undoubtedly learned discipline under the pressure of a leader who set demanding goals that on many occasions put his tanks on the verge of running on fumes. Returning home after the war, Hess was back in the family business as it grew to meet America's expanding fuel oil demand in the 1950s. In the early 1960s, Leon was the CEO of Hess Oil and Chemical. The company now had the Port Reading Refinery located in Perth Amboy and Woodbridge, New Jersey, as well as multiple fuel storage tanks and an extensive commercial fuel oil delivery service.

In 1966, the breakout opportunity for Hess occurred when he seized the chance to purchase a 10 percent block of shares in the Amerada Petroleum Corporation for $100 million from the Bank of England. The bank had acquired this holding as part of Germany's World War II reparation payments. With this block of stock in hand, Hess Oil and Chemical in 1969 merged into Amerada Petroleum. In this transaction, Hess prevailed over a competing offer from Phillips Petroleum, and this success catapulted the company into an entirely new league. This acquisition provided Hess the ability to build a large refinery on the island of St. Croix in the U.S. Virgin Islands to process its anticipated share of production from the newly discovered Prudhoe Bay oil field on Alaska's North Slope. The company planned to bring the oil from Alaska through the Panama Canal to the Caribbean, capitalizing on an exemption that St. Croix enjoyed from the requirement to use U.S. flag vessels. This cost advantage amounted to $2.15 per barrel. In addition, he had the ability as a larger public company to use Amerada's high-quality production base in the Williston basin of North Dakota, the Gulf of Mexico, and the Permian basin of West Texas to fund what would become a major new source of profitable offshore production in the United Kingdom (U.K.) and Norwegian sectors of the North Sea.

All of that progress lay ahead, but on this first visit I learned how plans can, and sometimes do, go very wrong with big new energy projects. In this case, the fact emerged during the New York City meeting that the company's Virgin Island refinery had a problem that would significantly affect its near- and intermediate-term earnings outlook. By 1971, it became clear that regulatory and environmental delays in developing the North Slope reserves meant that imported oil from Africa, Venezuela, and Middle East sources would be needed at the refinery. Unfortunately, the supertankers, or very large crude carriers (VLCCs), intended to bring imported oil to the facility from the new far-off sources drew too much water to be unloaded. Furthermore, the usual solution of dredging a deeper channel to the refinery dock was not a viable option, because the coral around the island could not be dredged. As a result, the startup and running of the refinery to full capacity would have to be postponed pending the development of a more expensive and time-consuming alternative of an oil lightering unloading plan. Earnings of $100 million in 1971 plummeted to only $15 million the following year. Since the Amerada Hess stock price had recently almost tripled because of growing demand for its fuel oil and the prospect for even more growth as the new refinery came online, only three months into my analyst career I experienced my first case of "missed corporate expectations." During the prior year, the stock had moved from $25 to over $70 per share. At the time of the analyst meeting, it had settled back into the mid-$50s per share. Three or so days after the meeting, the share price dropped by over 25 percent and ultimately completed a round trip back to about $25 per share. Notwithstanding this telling lesson about the harshness of short-term market judgments when disappointment versus investor expectations is manifest, I was fascinated by the evolution of Amerada Hess into a case study of an American entrepreneurial business endeavor. Over many years of observing and analyzing the ups (and downs) of strategic moves made by Leon Hess and subsequently those made by his son, John, the company has evolved into a top Fortune 100 company, now renamed the Hess Corporation. I have come to appreciate how this company assembled attractive upstream exposures internationally, as well as in the domestic unconventional resource arena, with strong positions in the Bakken oil development of North Dakota and the Marcellus and Utica plays of Appalachia.

Despite his successes, Leon Hess never let himself forget his humble beginnings in business. In the late 1970s, while proudly giving a group of analysts a tour of his spotlessly clean St. Croix facility, he took several of us to the garage for fire trucks to show us the fuel delivery vehicle he personally drove during the Great Depression. Sometime in the early 1970s, he had moved it to St. Croix and caringly reconfigured it as a supplemental fire truck. It clearly evoked treasured memories of his early years building the business.

My next visit was to Harry Shuttlesworth, the treasurer of Union Pacific Corporation (UNP). While this company was one of the original transcontinental railroads, it had also recently acquired a large oil operation called Champlin Petroleum that provided a Texas Gulf Coast refinery, which complemented the company's seven million acres of oil and gas prospective land grant holdings that had been awarded by the U.S. government when the railroad was initially built from Omaha to Utah just after the Civil War. UNP's management viewed the Champlin transaction as a way to exploit more fully what had previously been its relatively passive petroleum mineral holdings.

Colonial Management held a significant position in Union Pacific stock, and I was invited to go on an analysts trip to Omaha and then on to Denver and Corpus Christi to learn about UNP's natural resource activities. In many ways, this trip ignited my passion for the energy sector. I had no idea at the time about the myriad future events for which this introduction to UNP and the Rocky Mountain oil industry would prepare me. As it turned out, there were many more such occasions than I could imagine. This particular trip deepened my understanding of the company's extensive "land grant" holdings. These were a vision of Abraham Lincoln's, as an incentive for private capital to link by rail the eastern states with the western regions and territories.

The historian Stephen E. Ambrose has recounted in *Nothing Like It in the World* that the "land grant awards" were casebook examples of highly economically effective government incentives. The program's execution had numerous well publicized flaws in terms of inefficiency, fraud, and waste, as other historians have chronicled in detail. Nevertheless, the award of checkerboarded land (alternating square-mile sections) for twenty miles on either side of each mile of track installed was a powerful incentive for capitalists to finance and build the railroad on a timely basis under remarkably challenging conditions.

Ambrose acknowledges that the land itself had limited value in terms of the utility of the surface (even for cattle ranches, because of the sparse and inadequate grass and water). However, the accompanying rights to the subsurface minerals, including coal, iron, and petroleum resources, would over time become very significant. These holdings totaled some seven million acres across what eventually became four states and covered large portions of more than a half-dozen distinct geologic basins. These were variously prospective for oil, gas, coal, uranium, and trona. As Ambrose has observed, this development project employed many thousands of former Union and Confederate army veterans in the construction of more than an eight-hundred-mile railroad right-of-way. As a result, it had the powerful ancillary benefit of assuaging North versus South tensions in the early years of post–Civil War America by refocusing the nation on connecting the rapidly urbanizing eastern United States with the still open and rural West.

This trip also began to acquaint me with the potential merits of basing my

THE UNION PACIFIC LAND GRANT

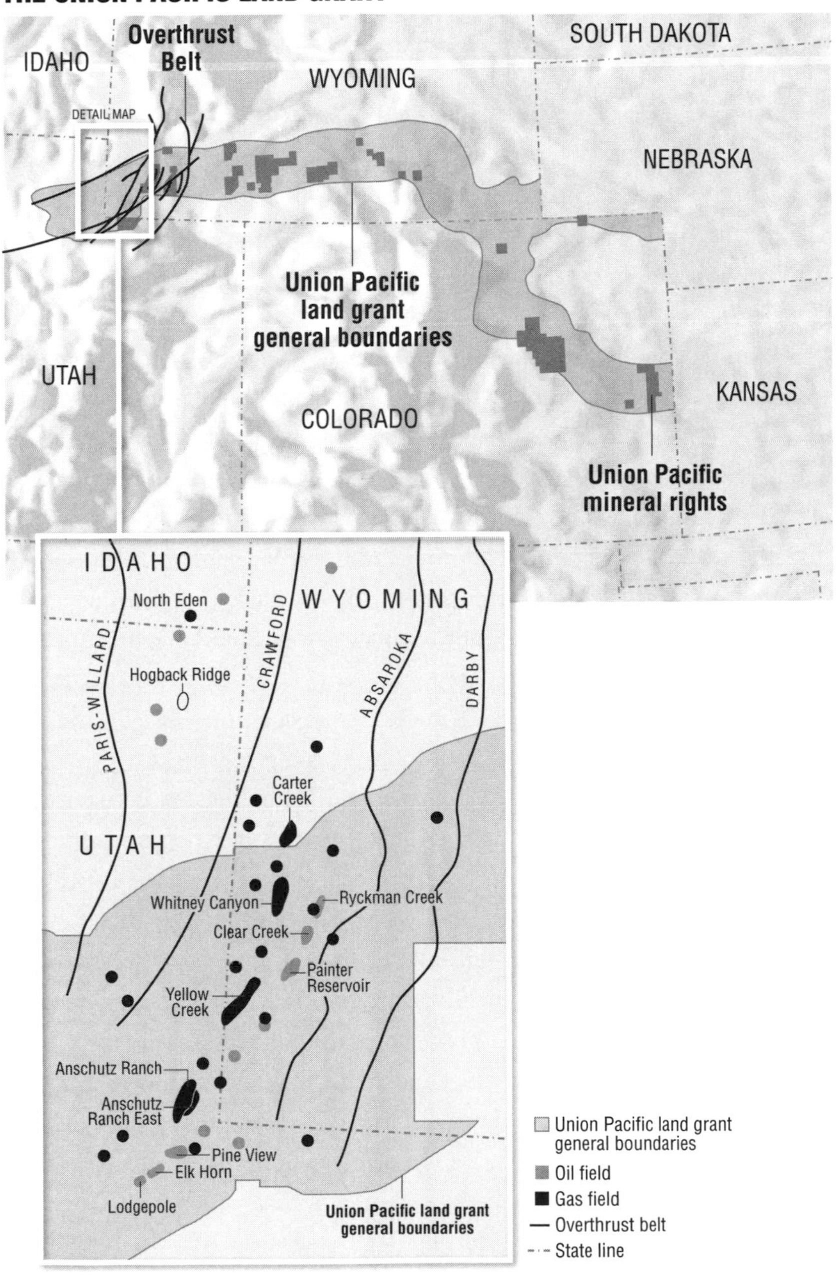

career and business efforts in the West (more specifically, in Denver, Colorado). Accomplishing that move would take almost a full decade, but I owe many thanks to Harry Shuttlesworth and the Union Pacific Corporation for opening my eyes to career possibilities in the American West.

Another episode in my early development as an oil analyst involved a first trip to meet T. Boone Pickens at Mesa Petroleum in Amarillo, Texas. Little did I realize then that this iconoclastic petroleum independent would do as much to affect the beginning of my later career as an investment banker as just about anyone. Then, however, my lesson from the early 1970s Amarillo trip was that Mesa Petroleum's investment in the cattle feeding business was a distraction for Mesa's management from its core focus on the oil and gas business. I was still relatively inexperienced, so my conclusions about the company were tentative at best. My instinct was to err on the side of caution. Colonial did not have a position in the company at that point, and I did not recommend initiating one. However, I did find Boone to be an engaging and strategically focused player in the business, one well worth watching. That early judgment proved correct. In subsequent years, our paths crossed multiple times. In the 1980s, Boone became a leading corporate activist and launched unsolicited offers for numerous companies. Some ten years later, my First Boston colleagues and I became advisors to many of those companies.

Not long after meeting Boone, I met Fred Hamilton, who was visiting Boston in the process of taking Hamilton Brothers Exploration Company public. Hamilton was a pioneering explorer for oil in the United Kingdom's North Sea. In fact, some four years later, his company's development of the Argyle field would prove to be the first offshore oil brought onstream in the U.K. sector of the North Sea. This was my introduction to the opening of an exciting new world-scale oil province. It made me aware of how a relatively small but focused and creative independent oil enterprise could compete effectively against much-larger and more-established companies. It also motivated me to commit to ongoing research and analysis on the full range of players that would ultimately find and develop some six million barrels of daily North Sea oil production. For well over a decade, the North Sea became the largest new oil trend being developed in the world. For perspective, this level of production would ultimately amount to almost 10 percent of global output and exceed that of Kuwait and several other Middle Eastern and North African oil exporters. Following the Hamilton Brothers entities as a research analyst periodically provided a reason to visit the company's headquarters in Denver; in time, I came to realize that there was a compelling case to make that city my home.

Another of my early encounters with oil and gas enterprises involved Atlantic Richfield (ARCO) and the Prudhoe Bay field. During summer leave after graduation from West Point and before starting Ranger School, I worked as an automobile salesman in Suitland, Maryland. While I had some success at selling cars during those two months, there were times when activity was slow, and I would use the opportunity to catch up on what was happening in the business world. During one such occasion, I became aware of Atlantic Richfield's plans to drill an exploratory well in hopes of discovering a new potentially large oil field north of the Arctic Circle in Alaska. On March 12, 1968, ARCO and BP announced that the well was highly successful. Now some five years later, while working at Colonial Management, I began to get my arms around what all this meant in economic terms. In fact, to accomplish that task fully would take the better part of two decades, but it began in early 1972 with another trip to an analysts presentation at the St. Regis Hotel in New York City to hear Atlantic Richfield outline its plans to develop this world-class oil field. Equally important, I came to know Robert O. Anderson, the founder of ARCO, as one of the high-profile leaders in the industry in the 1970s.

By the end of my first year at Colonial, I felt that I had put in place an initial framework for analyzing the oil and gas industry. In addition to a series of company management visits, I had met several times with each of the leading Wall Street oil analysts who over time became instrumental in helping me develop a better understanding of industry fundamentals, as well as potential investment upsides and pitfalls. These included the late Barry Good of Laird & Co. (and subsequently Morgan Stanley), Charlie Maxwell (C.J. Lawrence), the late Dillard Spriggs (Baker, Weeks), Frank Reinhardt and the late Al Anton (both Carl H. Pforzheimer), the late Bob LeVine (Wertheim & Co. and later J.P. Morgan), the late Sal Ilacqua (L.F. Rothschild), and Peter Gough (First Boston). I shall always be indebted to them for their mentorship and friendship.

As I gained confidence and a degree of credibility with the portfolio managers of Colonial's various funds, I began to realize opportunities to influence the firm's energy investments. In particular, having developed a point of view on the giant Prudhoe Bay field in Alaska, I recommended additions to the Colonial Fund's holdings of Atlantic Richfield as the upside potential of developing the largest field ever discovered in North America became clearer. I also came to appreciate the extent of exciting new oil discoveries in the U.K. and Norwegian sectors of the North Sea, and I recommended Phillips Petroleum for certain funds because of its "game changer" discovery and development of the giant Ekofisk oil field offshore Norway. As the geopolitical clouds overhanging oil stocks started to lift, these companies' stock prices began to appreciate, and my confidence in making investment judgments based on improving economic fundamentals also strengthened.

About this time, another event occurred that significantly affected my future and necessitated a career decision. Over the Fourth of July holiday of 1972, Frank Tse, Colonial's thirty-two-year-old technology and office equipment analyst, was playing a game of pickup basketball; tragically, he suffered a heart attack, collapsed, and died. After his memorial service, I was asked to take on his industry coverage responsibilities while simultaneously training a proposed new hire to be my replacement as the firm's oil analyst. My reaction to this proposal was at best hesitant. While my West Point education was concentrated on electrical and other engineering courses as well as mathematics, I was not convinced that I wanted to focus on the technology sector itself. In part, I realized that I already had ample technology exposure within the energy industry I was covering. In contrast, Colonial Management's plan for my development called for me to rotate over time as an analyst through several different industry sectors, all as preparation to become a portfolio manager. I was not at all sure that this was the path I wanted to pursue. In addition, because my first year had ignited a genuine passion for analyzing the petroleum industry with all its technical complexities, intriguing personalities, and complex geopolitical issues, I intuitively felt a deep desire to stay the course with the energy sector that was already so strongly resonating for me.

Most fortuitously, an opportunity to do just that developed while I was having these discussions with Colonial. H.C. Wainwright & Co., a hundred-year-old member of the New York Stock Exchange and a firm with an outstanding reputation as a purveyor of in-depth institutional investment research, approached me indicating that they were looking for someone to be part of their oil research team that was headed by Bob Meyer. The desire to stay on my chosen path prevailed. In early 1973, I moved to New York City to join the firm where I would complete much of my development as a petroleum research analyst focusing on fundamental and strategic positioning within the oil and gas sector.

The Yom Kippur War, October 1973

Markets work but do so in their own time
and at their own pace.

In early October 1973 as a new "sell side" institutional research analyst at H.C. Wainwright, I was attending an oil conference organized by the National Association of Petroleum Investment Analysts (NAPIA) at the Arizona Biltmore Hotel in Scottsdale. During the conference, there was a news announcement reminiscent of my previously described USMA graduation parade experience. It declared that war had again broken out between Israel and the surrounding Arab states. A decision by President Nixon to provide Israel with replacement military equipment and supplies greatly upset Saudi Arabia's King Faisal. This commitment caused him along with other Arab oil-exporting countries to initiate an oil export embargo against the United States and selected European countries. I was in London a few weeks later when that decision was actually being implemented by the Arab oil-exporting countries that were members of OPEC.

This was a fascinating situation in which we saw a firsthand demonstration that some markets do *not* always (or at least not immediately) behave rationally. Rather, they sometimes can be irrational in the most counterintuitive and even perverse ways, at least for a while. When this occurs, it is often because market decision makers either learned the wrong lesson from an earlier experience or at least drew the wrong conclusion about how to apply the lessons learned from an earlier experience. This particular situation once again involved the oil tanker market. As mentioned previously, in the Six-Day War of 1967, the closing of the Suez Canal had caused tanker rates to skyrocket. This event resulted from the much-lengthened journey to take oil around Africa to Atlantic basin markets. The fixed number of relatively small ships had dramatically shrunk the total shipping capacity of the global oil transportation system. Thus there was a bidding war for the limited number of ships available to haul the oil needed in consuming markets.

However, by late 1973, the situation was substantially different from that of 1967. The building and launching of a large fleet of supertankers in response to the increase in tanker rates in the late 1960s had actually created more oil carrying capacity than was needed worldwide. Furthermore, the Arab embargo at least threatened for a while to reduce the amount of Middle East oil to be carried to these markets. Accordingly, the logical conclusion from the fundamental changes in the supply/demand relationship indicated that tanker rates should fall much further. They were already somewhat off their previous highs because of the oversupply of ships, but for a while at least, just the opposite was actually happening! Tanker rates escalated sharply, much as they had in the 1967 war, as apparently many market players assumed that history was about to repeat itself.

I became aware of this anomaly on a mid-October visit to Mobil Oil's London office, where I met one of that company's senior managers of worldwide shipping. He remarked to me how perplexed he was that the markets were behaving so differently from what he considered the supply/demand fundamentals indicated should occur. He also told me that Mobil was inclined to sell forward almost all of its shipping capacity at these much-elevated rates, because they could not imagine these levels being sustained. This was not a momentary market aberration. Instead, this incongruous situation lasted for many weeks, enabling Mobil to complete its large sale program before the overall market embraced the reality of the changed supply/demand dynamics, causing tanker rates to collapse around year-end. As a result, for the next several years Mobil enjoyed a significant transportation cost advantage over many of its competitors. Rather than analyzing the full range of operative fundamentals, many other companies had taken actions based on historic market behavior in an inappropriate (or at least not an applicable) precedent.

When Correlations Fail

The risk of looking backward.

In subsequent years, I have observed numerous other cases of this same type of misjudgment about the applicability of historical precedents in other phases of the energy and securities marketplaces. A piece of wisdom often attributed to Mark Twain asserts, "History doesn't repeat itself—at best it sometimes rhymes." As an extension of that analogy, it is important to note that words that sound alike can be very different, and the difference is what really matters. Many backward-looking computer trading models are susceptible to this same problem. When they are based on historic correlations, such models can and sometimes do fail to take account of one or more key differentiating factors that make the presumed historic analogy (or analogies) not nearly as comparable as first presumed. In the securities and commodities trading arenas, this type of risk is often overlooked and was thus periodically the source of systemic capital destruction, much as it was in the example involving Mobil's counter-parties in the oil tanker transactions for much of the balance of the 1970s.

In more recent periods, this same tendency to become overly dependent on backward-looking (that is, presumed historically analogous) situations for guidance about how the future may unfold has often been a losing strategy. In fact, it has been likened to driving forward while looking in the rearview mirror to avoid obstacles, a clear prescription for trouble. In many respects, the late 1990s failure of Long-Term Capital Management (LTCM) stemmed from the use of high financial leverage along with this same defective, backward-looking approach to investment analysis. Put another way, presumed historical analogs often work until they do not. Then they can lead to disastrous consequences. Usually the cause stems from the emergence of some (often subtle) factor that differentiates the current situation from its apparent historical look-alike. Other potential examples from more recent past periods involved the 2006 failure of Amaranth Energy, a partnership set up to trade natural gas based on historical relationships, and, I suspect, the JPMorgan Chase "whale trade" hedge that failed to perform as expected in the second quarter of 2012. This episode gives new meaning to the term "a costly misjudgment," with recognized trading losses to date of over $6 billion, along with $920 million of related penalties.

In early 1975, I was most fortunate to have Paul Leibman join me as part of Wainwright's energy research team. He had previously been an equity analyst at the St. Paul Companies, an insurance company in St. Paul, Minnesota. Paul was a key addition to our petroleum sector research efforts. He brought impeccable analytical skills, a healthy skepticism, and a strong competitive desire to deliver top-quality, often somewhat contrarian investment insights to Wainwright equity research publications. I, as well as our firm's clients, benefited from his contributions to our team. I was also fortunate that we would ultimately work together in two more enterprises (First Boston and Petrie Parkman) over a total of seventeen years. Paul subsequently formed and successfully operated an energy hedge fund and now manages funds privately.

As a result of the much-expanded research capacity that Paul's addition brought, we began what proved to be a highly productive three-year program, during which we developed and published in-depth reports on a series of exciting industry developments in the upstream petroleum sector. These reports included multiple analyses of new world-class oil discoveries in the U.K. and Norwegian North Sea, a broad assessment of major producing fields throughout the United States with a view toward their potential for enhanced oil recovery projects, especially involving carbon dioxide floods. We also provided ongoing assessments of new exploratory developments including the Rocky Mountain Overthrust Belt, the Tuscaloosa Trend, the Smackover Norphlet Jurassic Trend, and offshore exploration in the Gulf of Mexico, Gulf of Alaska, Santa Barbara Channel, Santa Maria basin (offshore), Beaufort Sea, and eastern Canada's offshore Atlantic basin. Finally, we published specific reports on the dozen and half or so oil and gas companies in our ongoing research coverage.

Chapter 2

A TUMULTUOUS DECADE

Experimenting with Deeply Flawed Petroleum Regulation

The law of unintended consequences was powerful and ever present!

In August 1971, with inflation having risen from less than 2 percent and threatening to approach 4 percent, President Nixon embarked on a course of action to contain rising consumer prices. That ultimately futile effort would have ramifications well beyond the end of his presidency. As Richard H. K. Vietor describes in *Energy Policy in America since 1945*, "Congress, the Executive Branch and the courts devoted a ridiculous amount of time and political energy to the issue of price controls." These initial actions, in combination with Nixon's decision to take the United States off the gold standard (a de facto devaluation of the dollar), involved promulgating a series of rules to limit price increases for a variety of goods, services, and commodities. At the time, these measures seemed fairly moderate. Early on, they were even judged to be effective in some quarters in that the still relatively mild inflationary pressures initially showed hints of diminishing. However, in the aftermath of the Yom Kippur War, the role of oil price–driven inflation began to escalate and triggered knock-on effects throughout the supply chain for other forms of energy, as well as for the cost of manufactured goods and key consumer services. A widening range of businesses were soon forced to pass on these higher costs in the products they were manufacturing and services they were selling. Ultimately, this rising cost structure resulted in higher labor cost demands and marked the onset of a full-blown, self-reinforcing inflationary cycle.

With the U.S. dollar no longer convertible at a fixed rate into gold, the cost of imported goods began to rise. I saw this personally beginning in 1973–74.

On visits to see investment clients in West Germany, I noted that the old conversion rate of four deutsche marks to the dollar that I had enjoyed as a U.S. Army officer in the late 1960s was now two marks to the dollar. Put another way, the old deutsche mark that once was worth a quarter now commanded fifty U.S. cents. In turn, this sharp change in the currency exchange rate had obvious inflationary implications for the cost of growing amounts of German manufactured goods imported to the United States. There was the unfortunate combination of lingering effects of the Johnson administration's ill-advised deficit financing of the Vietnam War along with a weakening dollar. These adverse trends and the pass-through of much higher oil prices began an inflationary spiral that by mid-decade was beginning to feed on itself. By 1978, inflation in the United States would be running at 9 percent, entailing particularly adverse implications for credit markets.

As Vietor has detailed, crude oil price controls were initially instituted in August 1971 as part of an economy-wide price freeze. This was phase 1 of the economic stabilization plan. During phase 2, which involved the period November 1971 to January 1973, there were controlled price increases to reflect cost increases with profit limitations. Phase 3, which covered the period January 1973 to August that year, provided for voluntary increases of up to 1.5 percent annually for cost increases. This period was then modified by Special Rule No. 1 during the period March 1973 to August that year, which imposed mandatory controls for the twenty-three largest oil companies. Phase 4 instituted two-tier pricing with old oil set at the level of May 15, 1973, plus thirty-five cents while new oil, stripper oil, and "released" oil were decontrolled.

The measures put in place beginning in 1973 to contain oil price increases became more problematic as the decade unfolded. Late that year, William Simon, a former Wall Street bond trader from Salomon Brothers, was installed as head of the newly created Federal Energy Administration and was designated as the nation's first energy czar. The Watergate scandal culminated in a change of leadership with President Nixon's forced resignation on August 9, 1974. His successor, Gerald Ford, then attempted to contain inflation by launching the ill-fated Whip Inflation Now (WIN) campaign.

To try to understand better where the U.S. regulatory initiative in energy matters was headed, I made a trip to Washington, D.C., and met with Mr.

Simon. I found him to be a strong personality, very focused and confident of his decision-making abilities. His style communicated quickly that he would not suffer fools for long, arguably a positive in dealing with entrenched governmental bureaucrats. A couple of decades later, in the early 1990s, having achieved noteworthy success as a private equity investor, he and his partner, former First Boston Corporation chairman Al Shoemaker, became an early client of Petrie Parkman & Co. Our paths crossed one more time in the late 1990s when he endowed the Simon Center for Professional Military Ethic at the U.S. Military Academy at West Point. Having also served capably as Secretary of the Treasury in the Ford administration, Simon was a proven leader, an accomplished executive, and later a successful private capital investor. However, despite his many talents as well as those of his able successor as energy czar, Frank Zarb, the effectiveness of these individuals and others as regulators of the energy sector proved to be elusive. In the end, the U.S. regulatory experience was counterproductive for both of these leaders and the nation.

The foregoing is instructive for those who would believe that effective price regulation in the name of consumer protection is largely a matter of ensuring the competence of personnel. As J. Howard Marshall has detailed in his autobiography, *Done in Oil,* the complexities attendant on daily market-driven decisions too often defy centralized and generic controls even by the most competent of regulators. Marshall had been an oil industry regulator in President Roosevelt's Department of the Interior during the economically challenging depression years of the middle 1930s. He subsequently was a senior executive at Standard Oil of California (now Chevron). He returned to governmental regulation of the oil industry again as a deputy to Secretary of the Interior Harold Ickes during World War II. Marshall then took on a series of successful senior management roles at Ashland Oil & Refining, Signal Oil and Gas, and Union Texas Petroleum. With this wide-ranging regulatory and private sector experience, he brought a unique perspective to the U.S. regulatory response to the first oil crisis of the early 1970s. When I spent time with him at California's Bohemian Grove in the summer of 1983, he underscored the deep reservations he had during the 1970s based on his earlier personal experiences involving the U.S. government's efforts to administer oil price controls. In that case, the goal was to try to put a floor under the price of oil that surplus production had driven from one dollar per barrel to as low as ten cents per barrel. In contrast, the U.S. government's regulatory efforts in the 1970s were to contain the upward price spiral. However, in Marshall's view, both challenges arose from the inherent complexities of regulating a commodity with an incredibly wide variety of quality, geographical differences, extraction costs, and utility to different consumers.

By the mid-1970s, the federal government had embarked on a multiyear process of price regulatory efforts intended to protect consumers. Unfortunately, these measures created a whole series of dislocations, misallocations, and distorting economic decisions that were highly constraining on the normal functioning of markets for oil, gasoline, diesel, heating oil, jet fuel, natural gas, petrochemicals, and many other downstream uses of these commodities. Vietor has noted, "The price of petroleum products was also regulated for most of that time (August 1971–January 1981), and quantities were allocated right down to the level of retail distribution." It became clear to me and other analysts that the "law of unintended consequences" was alive and well—in fact, even flourishing—throughout each of the U.S. energy equity markets and commodities exchanges engaged in trading petroleum products.

In the oil sector, a more than tripling in the world oil price in 1973–74 gave rise to calls in Washington for Congress to institute price controls to eliminate the "windfall profits" that were accruing to U.S. producers of oil, associated with higher prices for fields that had already been discovered and developed. This became known as "old" (or lower-tier) oil. Its price was, in effect, frozen at precrisis levels. Moreover, in time evidence accumulated that the price of some "old" oil needed to be allowed to rise at least moderately to ensure that rising future operating costs did not make its extraction uneconomic. That condition could lead to premature abandonment of older wells as well as further decreases in overall oil production. Accordingly, new rules were implemented to allow a limited degree of price escalation for "old" oil, as deemed "reasonable" by the regulators. As a result, important economic dislocations became manifest. Because large integrated oil companies tended to own and control more of the low-cost "old oil," they possessed a competitive advantage over independent refiners who were buying higher-priced imported and less-regulated "new oil." Attempting to address this issue in

November 1974, the government instituted the Old Oil Entitlements Program to equalize the cost of oil for all refiners.

In short order, the regulators had to acknowledge yet again the need for higher prices to incentivize the search for and development of new oil fields. Toward this end, President Ford proposed a new plan to decontrol the price of domestic oil. However, this initiative went nowhere with Congress controlled by the Democratic Party. In late 1975, responding to public criticism of the large increase in oil company profits at the time of unfolding economic stagnation, the U.S. Congress enacted the Energy Policy and Conservation Act (EPCA), which provided for a "windfall profits" tax. Layered on top of a system of multitiered pricing for chemically identical or functionally similar commodities, this tax and price control system created an incredibly complex set of issues in terms of effective audit and tax oversight. Starting in 1976 and extending well into 1978, the U.S. governmental regulators first in the Federal Energy Administration (FEA) and then in the subsequently created Department of Energy (DOE) were almost continuously pressed to issue new and ever more confusing (often ironically termed "clarifying") rules to deal with the unanticipated consequences of their prior rulemaking. A "new oil" category, designated as "upper-tier" oil, was given a higher but still controlled price, thereby creating a two-tier price system for essentially identical hydrocarbon molecules.

The situation became even more complicated as other oil categories were identified as needing incentives to develop (or sometimes just to perpetuate their continued production). For example, "lower-tier" oil that experienced enough of a production decline to qualify as "stripper oil" became an "approved" category priced at world oil market levels for U.S. wells producing less than ten barrels per day. Here the compelling case was made that continued regulation at an inherently arbitrary "old oil" price would force premature abandonment of production, constituting perhaps as much as 20 percent of total U.S. oil output. Another example justifying a world market price was the creation of a category for enhanced recovery oil for fields with governmentally approved projects to increase the ultimate recovery of oil by virtue of measures to stimulate additional production beyond that expected from primary or secondary recovery techniques. These tertiary technologies typically involved the injection of carbon dioxide or other chemical surfactants to flood oil reservoirs and thus sweep otherwise immobile residual oil in place from the field, or the use of steamfloods (also known as continuous steam injection) to reduce the viscosity in heavy oil fields of California to improve oil flows.

These numerous categories of oil that were subject to differing price treatment created progressively more complex (and ultimately convoluted) markets for domestic oil. For example, in my research analyst notes dated June 6, 1977, from a meeting with Ray Koeller, the investor relations manager of Standard Oil of Indiana (later Amoco), there is the reference to that company receiving the following realized prices per barrel: stripper oil, $13.58; upper tier, $11.07; lower tier, $5.29; and the blended average, $8.17. Because of these variances, there were powerful incentives to reclassify oil fields from tightly regulated lower-tier "old oil" to the various "new" categories qualifying for higher pricing treatment. Not surprisingly, as world oil prices escalated, the economic motivations for reclassifications became progressively more compelling, and the concern inevitably arose that some of the reclassifications could be unwarranted or even fraudulent.

Ultimately, the petroleum price regulatory regime became virtually undecipherable and bordered on collapsing from its own weight. This was an inevitable consequence that Howard Marshall had foreseen in his own attempts to construct an oil price regulatory framework back in the 1930s. In fact, after considering the problem, he and a colleague changed their minds about the merit of even trying to institute price controls. Eventually they were able to convince Secretary Ickes that they "didn't know enough to write a price-fixing order which would have any chance of operating effectively." After recognizing the seriousness of the petroleum pricing challenge, Marshall took pains to recount his lessons learned to in a March 22, 1979, letter to the man who was then the energy czar, Secretary of Energy James Schlesinger.

President Carter, in confronting these energy/oil challenges, managed to top his predecessor's unfortunate WIN campaign acronym by calling the evolving energy situation the "moral equivalent of war" (MEOW). Skeptics and political opponents of his leadership quickly observed that the MEOW acronym had the roar of a pussycat. In April 1979, in recognition of the futility of oil price regulation as a mechanism for providing consumer protection, President Carter announced his decision to pursue a gradual (thirty-month) phaseout of oil price controls and the associated "windfall profits" tax. Ironically, in one of those twisted cases of historical injustice, while Carter received little or no credit for this initiative, President Reagan garnered essentially the full credit stemming from his inauguration day decision to immediately eliminate all that then remained of oil price controls. While Marshall indicates that his warning to Secretary Schlesinger had gone unheeded, the timing of the Carter administration's switch to a price decontrol mode and Vietor's accounts of events leading up to it actually suggest to me that there was only a small delay due to the need for consensus building to embrace the wisdom of Marshall's advice.

As the later years of the 1970s unfolded, it became clear that the problems associated with what proved to be futile attempts to contain inflation with rigid price regulations were not confined to oil. Natural gas also became an area of increasing and often counterproductive regulation. Early in the industry's history, natural gas was an unloved and unsought by-product of oil exploration and development. In the early 1970s, on my first trip to Calgary, I can still remember seeing natural gas being flared at wellheads near the city's airport because of the lack of a worthwhile market for the substance. Throughout North America, some natural gas was associated with oil production, having come out of solution in the oil itself as pressures dropped when oil came to the surface. In other cases, the search for oil actually resulted in discoveries of "dry" natural gas. For these "discoveries," the connection to markets was either nonexistent or very limited. This remained the case until after World War II, when there was a resumed focus on construction of long-distance pipeline systems in the 1950s and 1960s. These projects had been essentially halted during the depression years of the 1930s and the war years of the early to mid 1940s. New and expanding pipelines began to link natural gas accumulations in the Gulf Coast, the Southwest, and the West to much larger consuming markets in the North and East.

In a 1954 watershed case involving Phillips Petroleum, the U.S. Supreme Court addressed a longstanding ambiguity in defining congressional intent regarding the scope of regulatory authority under the Natural Gas Act of 1938. The court ruled that the Natural Gas Act provided the federal government the authority to control prices over the entire natural gas industry, including companies that produced natural gas for sale to interstate pipelines. Thus, this decision clarified that such producers were subject to federal regulation of natural gas prices. The Federal Power Commission (FPC) became the vehicle for exercising this regulatory control. To protect consumers, the latter agency decided to hold prices at low levels, which sometimes translated into prices of only fifteen to thirty cents per million British thermal units (BTUs). Given the lingering surplus supplies of natural gas that had been found during the boom years of U.S. oil exploration in the 1920s and 1930s, many of these prices still prevailed in sales contracts even into the early 1970s. However, as U.S. markets developed for cheap and relatively clean natural gas in the early to mid 1970s, the gas surplus overhanging the market finally began to dry up.

As Vietor describes, early evidence of this change actually began to appear in the late 1960s. It became more broadly evident in the early 1970s, and by the mid to late 1970s the production shortfalls had mushroomed into large curtailments of contractual deliveries of natural gas to interstate pipelines. What had been manageable shortfalls of 286 million to 649 million cubic feet per day in 1971 and 1972, respectively, became much more challenging shortages of 2.975 billion to 3.77 billion cubic feet per day by mid-decade.

The concerns about natural gas supply reached a full crisis level in the winter of 1976–77. There was reduced gas production because of a combination of natural declines in mature fields and frozen wellheads in an unusually harsh winter. The increased heating demand stemming from these exceptionally cold temperatures gave rise to widespread news coverage of the shortage of natural gas to heat and power schools, factories, and homes. As Arlon Tussing and his colleagues have noted, "In the winter of 1976–77, curtailments cost the U.S. billions of dollars in industrial output as over 9,000 industrial plants were idled for lack of gas." They also noted that this resulted in three-quarters of a million layoffs and forced the closing of hundreds of schools. At this point, Congress enacted the Emergency Natural Gas Act to provide governmental authority to intervene for the short term in gas distribution decisions. These events contributed to something bordering on mass hysteria in the public's reaction to the country's potentially running out of natural gas in the near future. The shortfalls were very real, albeit short term, because of the extreme weather of that winter. The fundamental problem, however, was a basic lack of economic incentive to develop new supply. For several years, there had been only limited federal lease sale offerings in the Gulf of Mexico. Importantly, with natural gas still underpriced versus other energy sources, development of new onshore sources of gas was seriously lagging.

By 1974, curtailments to industrial customers amounted to 16 percent nationally. Furthermore, the development of new, higher-cost natural gas sources offshore the Gulf of Mexico in deeper water and ever deeper horizons onshore in Texas and Louisiana and in more remote parts of the Rocky Mountain West all gave rise to pressures to reconsider the price levels at which gas needed to be sold at the wellhead. Also, a two-tier price structure evolved. As the 1954 Supreme Court decision had made clear, gas developed and consumed in a single state was not subject to FPC regulation. For example, in Texas, Oklahoma, and Louisiana, for gas produced and consumed in those states, natural gas prices were a multiple of the interstate regulated price. Not surprisingly, gas development in these localized areas was flourishing in response to the higher prices. These regional economies benefited as their domestic gas became a cornerstone driver of broadened industrial growth. Over time, this example gave rise to calls for a nationwide initiative to decontrol natural gas prices.

In September 1977, responding to a Carter administration initiative for natural gas price deregulation, Senator Henry M. "Scoop" Jackson, chairman of the Committee on Energy and Natural Resources in the U.S. Senate, requested that the nonpartisan Congressional Budget Office (CBO), headed by Alice Rivlin, prepare a comparative analysis of natural gas pricing proposals. In his cover letter to the document that resulted, Jackson, who was well known for having coined the phrase "obscene profits" in referring to the profits of the oil companies in 1973–74, described natural gas pricing as "one of the most difficult public policy issues facing us today." He acknowledged that while there had been legislation pending before Congress regarding the gas price issue ever since the Supreme Court ruling of 1954, "no substantial modification to the Natural Gas Act was enacted" in those twenty-plus years. Over two decades of Congress failing to act on an important energy policy issue was an important contributing factor to the unfolding crisis in natural gas supply.

U.S. FOOTAGE DRILLED AND DISCOVERIES PER FOOT (1970–1976)

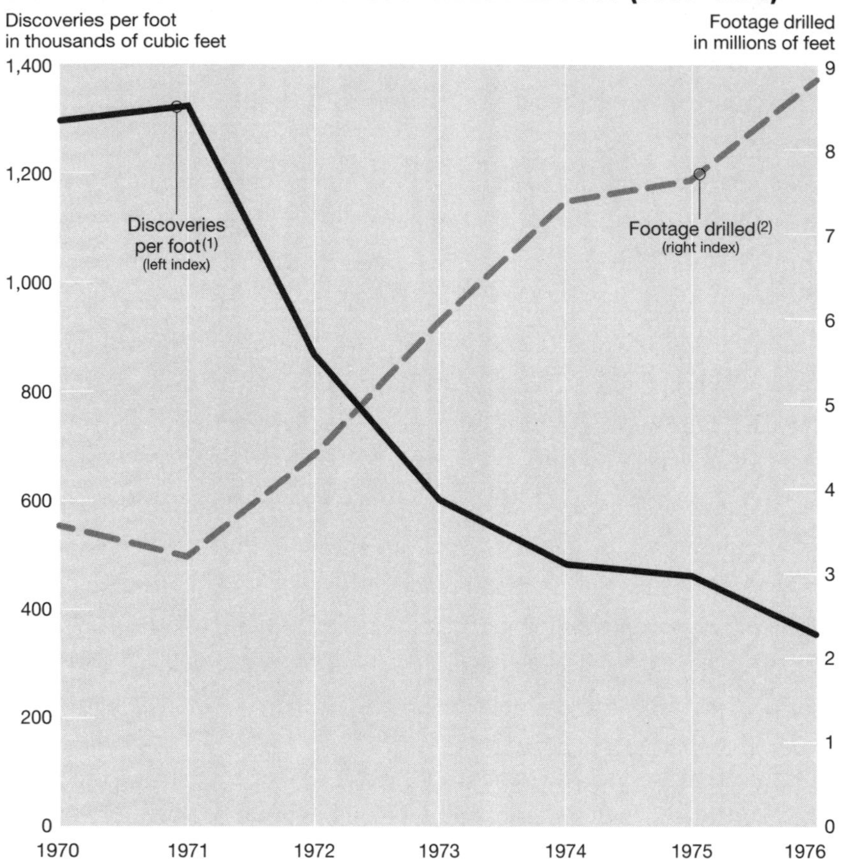

(1) Successful gas exploratory drilling footage.

(2) Total of new field discoveries and new reservoir discoveries divided by successful gas exploratory drilling footage.

SOURCES: Discoveries from AGA-API Reserve Report, American Petroleum Institute, May 1977.
Footage drilled from API *Quarterly Review of Drilling Statistics.*

The CBO report does attempt to present a balanced analysis of the pros and cons of deregulation. Unfortunately, its view of the benefits in terms of improved supply likely to occur under deregulation was heavily influenced by the disappointing trends in finding new natural gas fields over the period 1970 to 1976. The CBO noted that with more than a doubling of footage drilled, natural gas reserve additions per foot drilled declined by over 70 percent. This led the CBO to conclude that the supply response to be expected from higher deregulated gas prices would be small and that the economic burden on consumers would be significant. The report also postulated "alternative approaches" that could cushion the impact of price decontrol on consumers, provide incentives to conserve, and still use incremental pricing and tax incentives as a way to motivate companies to find and develop new sources of natural gas. Thus, this report paved the way for Congress to debate alternative courses of action for another seventeen months rather than directly confront the problem.

The growing negative psychology of the moment only brought out poorly constructed legislative responses by the U.S. Congress. The incentive was in place to find parties to blame. Congressional oversight hearings were the longstanding political mechanism predictably called upon for such efforts. These hearings provided the foundation for congressional passage of two pieces of legislation that had the effect of making a difficult situation even more complicated. Influenced by the 1977 CBO analysis, Congress passed and the president signed the Natural Gas Policy Act (NGPA) of 1978 in the fall of that year. The NGPA established nine price categories, with each having subcategories depending on the different sources of gas. This introduced administratively hopeless complexity for again essentially a chemically identical commodity. Furthermore, its companion, the Industrial Fuel Use Act of 1978, was predicated on the notion that the United States was running out of natural gas and that there was little or nothing that could be done to avoid the inevitable shortages. Thus, this bill was designed to set priorities for directly controlling the use of natural gas. This was an effort to deal with the presumed inevitability of progressively more onerous supply shortfalls. As Tussing and colleagues relate, "neither Washington nor the industry anticipated that gas would ever be in surplus."

The unfortunate consequence of the politicized and polarized atmosphere prevalent in the mid to late 1970s is that it obscured the ability of policy makers and the industry to appreciate and pursue options with real potential to effectively address the supply challenges. For example, the advances in "bright spot" geophysical analysis during the 1970s represented a major improvement in reducing the exploratory risk of developing new offshore sources of natural gas in the Gulf of Mexico. In effect, exploratory wells that once exhibited on the order of a one in six (that is, 16 percent) chance of success became transformed into opportunities with about a four in six (that is,

67 percent) chance of success. Similarly, in the prevailing skeptical environment, policy makers apparently underappreciated the enormous potential identified by some industry experts for tapping large quantities of natural gas in so-called tight reservoirs in many onshore basins of the western states. Nevertheless, techniques involving fracking and other reservoir stimulation measures were improving for unlocking such low-permeability fields. These sources played an important role in creating surprising gas surpluses in the 1980s and early to mid 1990s. In retrospect, it seems likely that an earlier and more flexible approach to natural gas pricing along with accelerated offshore lease sales could have substantially mitigated the forces contributing to the late 1970s crisis in natural gas.

Thus, the future turned out totally different than was widely expected in the 1970s, especially in governmental circles. Following price increases in the 1980s that incentivized new resource development, instead of chronic shortages dictating enforced rationing, there actually emerged more than a decade of chronic natural gas surpluses. However, considerable ongoing damage was done by the earlier legislative actions. These took more than a decade first to be discredited and ultimately to be discarded, as over time, the premises on which they were based became recognized as lacking merit or even attachment to reality.

An incident that occurred at a petroleum conference in 1977 in Washington, D.C., highlights my recollection of the tenor of the times. The conference was organized by the National Association of Petroleum Investment Analysts (NAPIA). In more "normal" times, NAPIA would meet in cities where oil and gas operations were an integral part of the economy. However, by 1977, the rule-making and policy initiatives by Washington regulators and legislators were dictating so much of the industry's operations that meeting at this center of energy economic and political power became imperative. The conference featured several speakers from key House and Senate energy committees who outlined their thinking on current policy matters and items of regulatory focus. At a late afternoon reception, I approached Louisiana's Senator Russell Long (now deceased) to ask him about a tax policy issue that had not been adequately addressed earlier that day. After hearing the query, the senator took one look at me and said, "Son, I don't know about that but I am on your side!" and then quickly walked away. That and other incidents have tended to make me a skeptic regarding the seriousness and professionalism of many (though thankfully, not all) congressional leaders in dealing with energy policy matters, despite the demonstrably critical role of energy policy in shaping U.S. economic effectiveness, especially in dealing with globalized markets.

From the middle of the 1970s onward, Paul Leibman and I had watched the petroleum sector morph into a hopelessly complicated and sometimes dysfunctional array of enterprises because of the changing cost characteristics of petroleum resource development as well as the progressively more highly regulated aspects of the sector. In looking back over those years, I now realize how much the oil industry's credibility and political effectiveness had become undermined by critics such as Christopher T. Rand, author of the book *Making Democracy Safe for Oil: Oilmen and the Islamic East*. Early on, I had not fully appreciated the widespread legacy of anti-oil sentiment that this book and others like it would generate. By the second half of the 1970s, selecting investments to recommend had inevitably devolved from assessing exploratory or operational fundamentals into exercises in identifying companies positioned to benefit (either by accident or design) from many of Washington's often poorly constructed policies and rules.

In the spring of 1975, the Securities and Exchange Commission's (SEC) ending of fixed commission payments in securities markets as a compensation mechanism for institutional investment research dramatically changed the economics for boutique research firms such as Wainwright. The full effect of this decision became evident over the next two years as most such firms merged into larger securities firms or went out of business. The hundred-plus-year-old firm in which I was a partner saw its revenues plummet by 75 percent in less than two years. A solid business model that had experienced stable growth for decades became obsolete in a matter of months. It proved to be a highly instructive personal reality lesson in the powerful changes that could and would be unleashed by decontrolled markets. Accordingly, I decided to join the First Boston Corporation as the company's senior oil analyst in late 1977, taking the place of Peter Gough, who was relocating to First Boston's London office. Paul Leibman joined Goldman Sachs about the same time.

In the aftermath of the first oil spike in 1973, world oil and securities markets became highly attuned to the potential for further Middle East conflicts

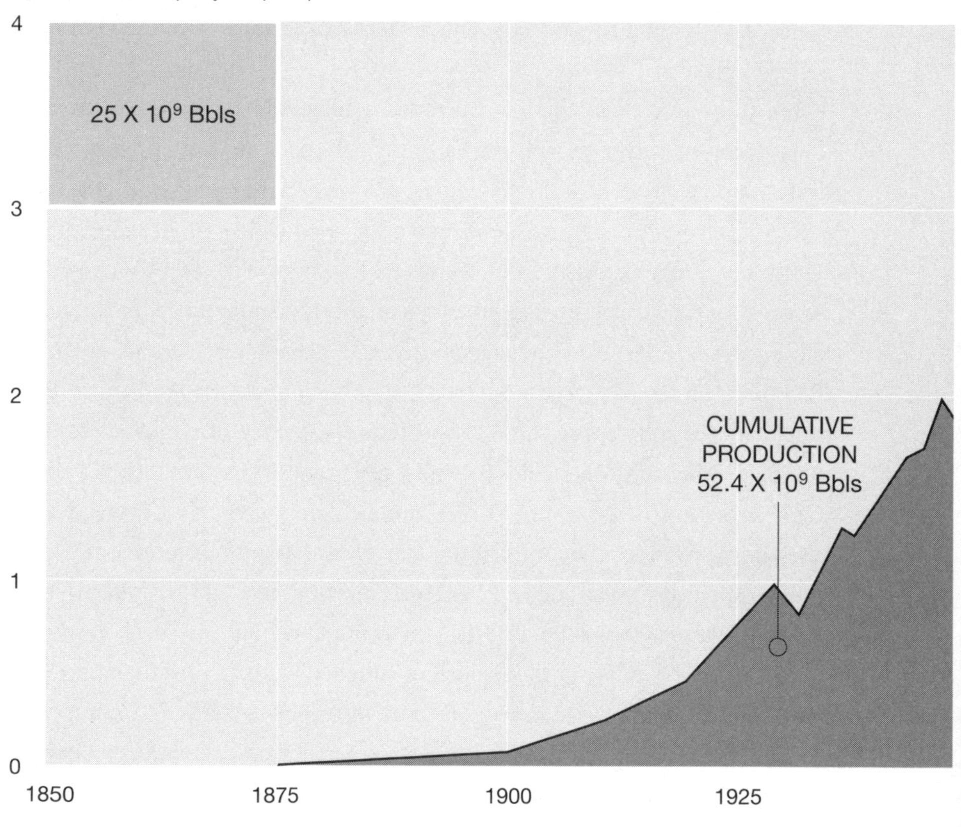

to disrupt oil supplies and wreak havoc on global oil pricing and Western developed economies. There was the broadening realization that a contrarian and controversial prediction in 1956 by M. King Hubbert, an ex-Shell petroleum geologist, had been correct and, in fact, even visionary. Hubbert, at that time an employee of the U.S. Geological Survey (USGS), had analyzed patterns of depletion for natural resources (focused on oil, in particular) on a basin-by-basin basis. His insight led him to the conclusion that the clearly observable symmetrical nature of cumulative field or basin depletion of petroleum over time would also apply to aggregated collections of hydrocarbon basins in the United States. Thus, his analysis of the maturity of resource depletion patterns evident in data available in the mid-1950s con-

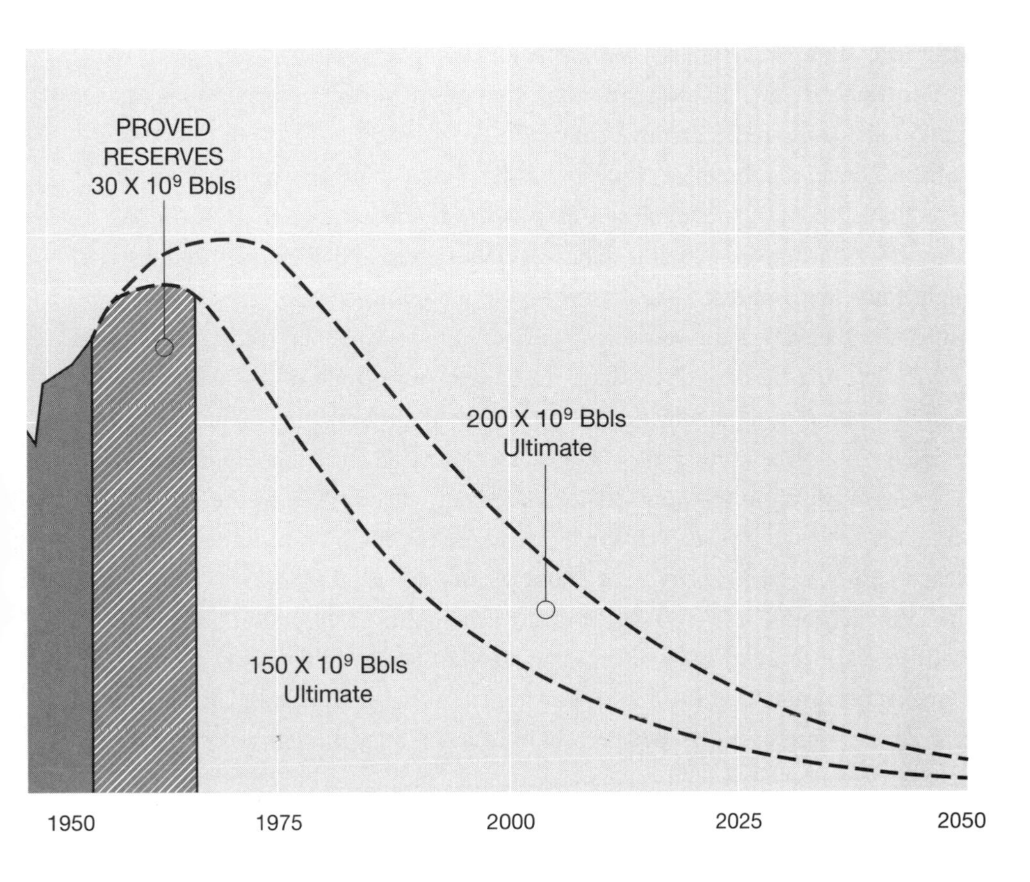

vinced Hubbert that total U.S. oil production was likely to peak once and for all some fourteen to fifteen years later, in 1969 or 1970.

Hubbert's prediction of the timing of this event was proven correct, though the largest ever single conventional discovery of American oil was yet to be made in 1968 at the Prudhoe Bay field in Alaska. Given that this new find represented the potential for about a 50 percent addition to total U.S. oil reserves, it is at least arguable that Hubbert's forecast might have been thrown off if the companies participating in the Prudhoe Bay discovery had been allowed to develop the field expeditiously. However, the physical magnitude of the project with long development lead times and the associated capital intensity indicated that, even then, Hubbert's prediction still held up remarkably well. As the 1970s unfolded, Hubbert was able to see his long-range forecast of persistent declines confirmed. Accordingly, he became a much-in-demand commentator on what might come next, especially in the aftermath of the oil-driven economic disruptions of 1973–74.

For over a decade following the confirmation of M. King Hubbert's prediction about the peak in conventional U.S. oil production, I continued to hold to a belief that he might still be wrong. I suspected that a combination of technological advances in oil recovery methods and an intensified exploratory effort in both new basins and deeper horizons of older basins could significantly expand reserve additions over the coming decade or two. The potential for enhanced oil recovery (EOR) techniques was then (and is even now) intriguing because of the large amount of remaining oil in place in existing reservoirs. Studies by the U.S. Department of Energy have identified residual oil in place in reservoirs suitable for commercially viable EOR projects as numbering in the hundreds of billions of barrels. Even in the 1970s and early to mid 1980s, the targets were generally thought and sometimes clearly known to be quite large. For example, cumulative oil production from the Permian basin in West Texas by 1980 was well over 20 billion barrels. With primary recovery rates of approximately 20 percent, this alone would suggest a remaining residual-oil-in-place target on the order of 80 billion barrels. Thus, an improvement of 10 percent in recovery rates through secondary or tertiary technology would mean production of another 10 billion barrels.

While the arithmetic of this expectation is sound, there were two problems in translating this notional opportunity into a timely reality. First, most EOR techniques are extremely capital-intensive and can often be logistically complex. Thus, the lead times for their implementation are unusually long. Second, these projects also often involve much higher operating costs. These considerations tend to cause companies pursuing such projects to be especially sensitive about taking on projects that could periodically be subject to break-even or worse economics in the typically volatile oil price environments as reflected in the price swings shown in the frontispiece. In practical terms, this means that the adoption of such projects tends to be spread out over longer periods than pure project analysis would indicate. Put another way, outside observers (myself included) tended throughout the 1980s to be too optimistic about how much and how quickly enhanced oil recovery could benefit U.S. production trends. In fact, the lesson of recent decades is that EOR projects have been helpful in mitigating the rate of production declines in existing oil fields, but they have not contributed to an actual reversal in overall U.S. production trends.

With respect to early optimism regarding additional conventional oil discoveries in new basins and deeper horizons, the record has been mixed but generally disappointing, with relatively few noteworthy exceptions. In particular, for most of the 1970s and well into the 1980s, there was a decided lack of encouraging U.S. discoveries. Drilling in the Gulf of Alaska, the Overthrust Belt of the Rocky Mountains, and the Smackover and Tuscaloosa Trends of the southeastern United States all fell well short of initial expectations. In sum, conventional oil discoveries in the giant and supergiant categories after the unlocking of the Prudhoe Bay field have proven highly elusive. The conventional oil field exceptions mostly involve the positive surprises post 1995 in the Deepwater Gulf of Mexico. Most recently, there have been encouraging results involving numerous unconventional "shale oil" (as opposed to "oil shale") opportunities onshore in the United States.

The difference between the terms "shale oil" and "oil shale" deserves a brief diversion at this point because of the much-improved outlook for the former and continuing questions about the latter. The term "shale oil" refers to fully thermally matured petroleum liquids locked in "tight," or low-permeability, shales in the earth's crust. Upon extraction, such oil is readily refineable into petroleum fuels such as gasoline and diesel. "Oil shale," in contrast, refers to a much less thermally mature substance known as kerogen, which requires considerable additional heating, or "retorting," to eventually become a refineable source of petroleum fuels. This distinction is important because during the 1970s and early 1980s, there remained pockets of misguided (or at least premature) optimism that the large identified oil shale resources of Colorado, Utah, and Wyoming believed to approximate the known reserves of Saudi Arabia would provide the United States with a strategic alternative to oil import dependency. We now realize that even with today's improving technology the problem of maturing kerogen into usable oil involves so much energy input that the net economics of oil shale production has, to date, remained stubbornly uncompelling. Such is not the case for shale oil, because of its advanced thermal maturity.

All in all, M. King Hubbert's prognosis on the timing of a U.S. peak in conventional oil production was remarkably accurate. Furthermore, as its prescience has become more fully appreciated and time-tested, Hubbert disciples have extended his work to focus on the inevitability of global peak oil.

By the latter part of the 1970s, Hubbert's prognosis contributed to an increasing awareness by investors and securities market makers of the importance of public disclosures relating to changing patterns of a company's oil reserves as a driver of intrinsic value. These public pressures gave rise to a decision by the SEC to convene an advisory board on oil and gas accounting. Upon learning of this initiative, I applied to become a member of the commission's advisory board to contribute a "user's perspective" on the types of improved disclosure that the SEC was seeking to require. David Norr, a certified public accountant by training, was a First Manhattan Corporation petroleum analyst and member of that board who encouraged me to participate. David had long advocated that there be better disclosure of information about oil and gas reserves so that the accounting for production operations would be made more transparent and thus more comparable from firm to firm. In particular, he was focused on improving the comparability of finding and development costs for oil and gas reserves and the profitability therefrom.

The SEC advisory board met roughly once a quarter over a period of about two years. In addition to those of us from the financial community, its membership also included representatives from several oil companies, as well as the accounting and petroleum engineering professions and a professor of accounting from a college in Texas. The SEC staff member overseeing our deliberations was Richard C. Adkerson, an accountant who later went on to become a partner at Arthur Anderson and is now the president and chief executive officer and a director of Freeport-McMoRan Copper & Gold.

The deliberations of the SEC advisory board became quite intense, in large part because there were many very thorny issues that needed to be addressed to improve the reserve and operating information disclosures. For example, Stanley P. Porter in conjunction with Arthur Young & Company presented a document entitled, "Highlight of 'A Study of the Subjectivity of

Reserve Estimates' and Its Relation to Financial Reporting." This and other input captured our reservations about the limits of reserve disclosure information. In the end, the SEC accepted much of our input and developed new rules for disclosure. While some of the issues of accounting comparability remained less than fully addressed with a determination by what became known as the Financial Accounting Standards Board rule 69 (FASB69), there was a substantial overall improvement in the degree of detailed information mandated for disclosure of proved reserves.

Perhaps the most important advancement was the requirement for a disclosure known as the standardized measure of discounted future revenues from proved oil and gas reserves. Just a few years later, the importance and analytical utility of this requirement became evident as we entered into a wave of mergers and acquisitions (M&A) transactions for oil and gas companies. My colleague and future partner, Jim Parkman, was among the first to recognize that these disclosures could be used to "back fit" a production profile for a company's proved reserves that could then be developed into a series of valuation scenarios for possible M&A transactions. During much of the 1980s and 1990s, Jim's development and refinement of this technique became a regular feature of our M&A advisory work whenever more-detailed internal engineering data was not available, and was particularly useful in the early stages of scoping out a potential transaction.

In retrospect, serving on the SEC advisory board provided lasting benefits over much of my subsequent career. The mandated disclosures became the industry standard for almost three decades. Ultimately, the need arose for even greater granularity of disclosures, as the technology for exploiting unconventional development of hydrocarbon resources advanced. This resulted in an SEC updated review in 2008 with new disclosure guidelines. All in all, I came to appreciate more about the governmental workings of Washington, D.C., in general, and the SEC, in particular. I also developed relationships with other board members that led to useful business insights as well as opportunities to work together again over the years that followed.

The Iranian Revolution and the Iran-Iraq War, 1979–1980

An anticipated cakewalk becomes an albatross, or Saddam Hussein encounters serious unintended consequences.

This second oil crisis provided an appropriate wrap-up to the tumultuous decade I have been describing. As the 1970s continued to unfold, securities markets became increasingly attuned to identifying and trying to anticipate the consequences of possible scenarios that might trigger the next "petro-crisis." One such speculative case involved the notion of the shah of Iran initiating military action to dominate his Arab neighbors in the Persian Gulf (or as Saudis and Kuwaitis much prefer to term it, the Arabian Gulf). In 1974, *New York Magazine* published a fictional account by noted novelist Paul Erdman of how such a strategic maneuver might occur, "The Oil War of 1976—How the Shah Won the World." Interestingly, this work of fiction was viewed as a credible scenario by at least one high-profile hedge fund manager.

However, development of this particular Iran-centric scenario was not to be the case. The next disruption did involve the shah, but in an altogether different sequence of events. What the author of the *New York Magazine* article, securities traders, and much of the rest of the world did not know was that by 1978, the shah would become terminally ill, and thus, he would be the initiator of the next disruption by virtue of his absence (or more correctly, his abdication) rather than by his overwhelming military presence. In the summer of 1978, the fact that he had been diagnosed with terminal leukemia was known only to a small inner circle, and it was a tightly held secret for another six months or so. In any case, the shah's rule in Iran was also no longer a popular one. His secret police (the SAVAK) had a well-deserved reputation for ruthless suppression of any opposition, using extreme techniques of torture and coercion. By the fall of 1978, pressure was building for the shah to liberalize or even abdicate, and his deteriorating medical condition made the latter choice the ultimate outcome.

By January 1979, the shah had been deposed, and in short order, what we now know as the Iranian Revolution was under way. The long-exiled Ayatollah Khomeini returned home to Iran from France. Almost immediately, senior Iranian military leaders, the last remnants of the shah's regime, were rounded up and summarily executed. For anyone doubting whether a change in rulers had occurred, the press accounts of the execution of the Iranian generals removed all doubt. The subsequent abduction and extended detention of hostages from the U.S. Embassy further solidified the message that previous American support of the shah's regime would be punished. The Shia Muslim clerics were now calling the shots, both figuratively and literally. The hostage crisis plagued the last two years of President Carter's term of office.

In September 1980, the Iranian Revolution triggered a reaction by its neighbor to the West. Iraq, under the Sunni Arab leadership of Saddam Hussein, initiated an invasion of an Iranian province to settle long-disputed territorial claims. This began the Iran-Iraq War, which would last eight years and result in over a million deaths and even more casualties. In launching the initial attack, Saddam apparently expected that his advantage of surprise would provide a quick and easy, even cheap, victory. This conclusion undoubtedly relied on Iran's disarray and demoralization stemming from the widespread clampdown on its population, including the diminished and devastated senior leadership of its army as the Iranian Revolution had taken hold over the prior year. While Iraq made gains early on, Saddam's army advances soon bogged down, and this conflict ultimately became known as the greatest World War I–type conflict (that is, trench warfare with all its horrors) since World War I.

Much as he would do repeatedly in future years in other wars, Saddam had both overestimated his own military capabilities and underestimated the will of his opponents. Saddam's misjudgments reflected his self-imposed isolation from realistic inputs from subordinates. Con Coughlin has labeled Saddam the "king of terror" because of his leadership by inducing fear, intimidation, and extreme coercion among his colleagues, staff, and Iraqi citizens. These measures were ultimately self-defeating in that they created an atmosphere in which the consequences of providing needed feedback and accurate negative information were so onerous as to motivate key personnel to tell the leader only the positive aspects of what he wanted to hear. In one especially noteworthy case, when dissatisfied with an idea broached by a cabinet member to help resolve the Iran-Iraq War, Saddam took the individual to a room adjacent to the meeting and summarily executed him. He then resumed his cabinet meeting and ensured that his act was widely understood as the likely consequence for any sign of perceived disloyalty.

As the war drew on for another seven-plus years, a terrible price was paid by both sides. In addition, world markets were deprived of five million barrels per day of oil for a period. This culminated in a second tripling of oil prices (from thirteen dollars to forty dollars per barrel) in less than a decade with broad adverse consequences for global economic growth.

With the benefit of 20/20 hindsight, it is clear that the significant impact on oil prices from the Iran-Iraq War played a key role in setting up an unprecedented wave of U.S. oil and gas mergers and acquisitions. These transactions resulted in the significant consolidation of companies in the U.S. petroleum industry that characterized the first half of the 1980s. The combination of the Iranian Revolution and the outbreak of war triggered the second tripling of oil prices in less than seven years with attendant large initial appreciation in publicly traded oil stock values. Yet however much the stock prices advanced, they still could not fully reflect the calculable discounted present value of the underlying oil reserves behind each publicly issued share. Moreover, once the negative impact on oil demand from much higher prices was triggered, petroleum prices and stock market values for oil equities entered a multiyear declining trend. Consequently, with the dashing of investors' much-elevated expectations for stock values, there began to develop rising shareholder pressures for corporate consolidations via mergers and acquisitions.

Chapter 3

REORGANIZATION AND CONSOLIDATION

My Move to the "Sellside"

Business model shifts necessitate a reassessment.

On November 7, 1977, I joined the First Boston Corporation to head up the firm's institutional research program focused on the petroleum sector. For the first couple of years, I devoted virtually all of my effort to establishing within the First Boston institutional platform the key attributes of the research franchise that Paul Leibman and I had developed at H.C. Wainwright (and its successor entity, Wainwright Securities). This effort involved analyzing, publishing, and marketing to major financial institutions a series of industry-oriented overviews as well as company-specific reports addressing the fundamental valuation parameters and strategic positioning of prominent oil and gas enterprises.

With this goal in mind, on the first day at work, I did not report to the firm's headquarters at 20 Exchange Place, just off Wall Street in lower Manhattan, but rather took a flight to Austin, Texas. There I attended a public hearing at which the Texas Railroad Commission was considering Marathon Oil Company's request to increase the allowed production from the giant Yates field in the Permian basin of West Texas from 100,000 barrels per day to 125,000 barrels per day. Back in 1975 as a Wainwright oil analyst, I had been to a similar hearing authorizing the field's production to double from 50,000 barrels per day. Thus, I knew this hearing would be informative but had no idea how helpful the insights gained from the trip would prove to be a short four years later when Marathon Oil merged with U.S. Steel. As on the prior occasion, the hearing was highly technical in nature and, in fact, amounted to a very compressed course in the matters of geology and petroleum engineering that combined to explain why the Yates field with its well over 3 billion barrels of oil in place was truly a world-class petroleum accumulation. What I learned during those four days of presentations, followed by a trip to the field itself south of Midland, Texas, has paid informational and investment judgment dividends time and again for over three decades.

This trip provided insights for the first of several equity research reports that helped establish my reputation as an analyst who could develop well-differentiated and insightful investment ideas based on a technical understanding of industry fundamentals. In turn, I used essentially the same analytical techniques over the next several years to develop similar investment research reports on Belridge Oil, Superior Oil, and General American Oil, all of which were ultimately acquired at meaningful premiums by larger companies.

By 1980 with much of this research foundation in place, I began my involvement in the broader scope of investment banking activity in First Boston's energy practice. A few years earlier, Joe Perella, who had joined the firm in the early 1970s, had convinced First Boston's CEO, George Shinn, to let him run a separately organized Merger and Acquisition Department to compete more effectively with Morgan Stanley and Goldman Sachs. Joe had also brought Bruce Wasserstein in from a leading law firm, Cravath, Swaine & Moore, to be co-head of M&A with him. Joe and Bruce made a formidable team, in large part because of their unique personalities and professional styles. They had skill sets (Joe, an accountant by training, and Bruce, a highly accomplished corporate attorney) that proved time and again to be highly complementary. Joe's gregariousness and engaging, upbeat curiosity about a wide variety of companies made him a natural and extraordinary rainmaker who consistently earned the trust of CEOs and CFOs, inspiring them to hire this upstart but still white-shoe firm, First Boston, as an M&A advisor. Bruce, a creative as well as focused strategic thinker, could consistently come up with out-of-the-box approaches to often-daunting merger and acquisition challenges.

Given the absolute size and ripeness of the petroleum sector for M&A-induced change, I was fortunate to be called on by these extraordinary individuals to help address the technical and financial complexities of our advisory assignments involving First Boston's long-established oil and gas clients. Initially, these projects involved DuPont (in its acquisition of Conoco), Marathon Oil, and Cities Service, but the list began to lengthen as the effectiveness of Joe's and Bruce's leadership caught the attention of clients that had historically been served by other firms, including Morgan Stanley, Goldman Sachs, and Lehman Brothers.

Relocating to Denver

An "oil patch" presence provides a competitive edge.

In early 1980, I concluded that the time was right to move to Denver, Colorado, where I was convinced that I could continue to pursue my career as an oil analyst while also raising my children in an area that was incredibly uplifting and affording of great business opportunity. After two years in Boston and almost seven in New York, I had built an extensive network of industry contacts and institutional relationships, as well as a credible record of research recommendations. Because Rotan Mosle was already an established investment banking and research boutique in Houston, I sensed that there was room for a Wall Street firm to have an exploration and production (E&P) investment research effort based in Denver and focused in part on Rocky Mountain petroleum activities in addition to the more global aspects of the oil and gas business.

The timing of such a move seemed propitious because of the excitement associated with a string of promising new discoveries occurring in an area called the Overthrust Belt of Utah and Wyoming, as well as exciting activity in numerous other Rocky Mountain petroleum regions such as the Williston, Powder River, and Piceance basins. Back in 1975, while researching upstream U.S. positions of major oil companies, I analyzed the discoveries of the Pineview field in Utah and the Ryckman Creek field about a hundred miles to the northeast in Wyoming, which opened the Overthrust Belt play. Several years later, I was on the First Boston trading floor when the discovery of the Anschutz Ranch East field was announced as a major upgrade to the size of the overall play. The initial well was drilled by Tom Brown, Inc., and a group of private oilmen, Ray Brownlie, Jim Wallace, Jerry Armstrong, and Joe Bander Exploration (BWAB). The principals of BWAB had moved to Denver in the late 1960s from Abilene, Texas, to focus on the petroleum basin of the Rocky Mountain states. The partnership of these four individuals epitomized the entrepreneurial spirit of private oilmen of that era in the petroleum sector. At various times they spearheaded the opening of new drilling plays in the Williston basin, the Michigan reef play, and a variety of other exploratory endeavors in Texas, Wyoming, and other western states of the Rocky Mountains. They often initiated new geologic plays and assembled substantial land positions. They then brought other independent E&P companies as well as major oil companies (such as Shell and Amoco) into the evaluation of the projects they initiated. When I arrived in Denver, they were highly regarded as leaders in the petroleum sector.

With respect to the Anschutz Ranch East discovery, several New York–based hedge funds had sold Tom Brown short in the belief that the stock was overvalued. On paper, their analysis seemed logical, but the market's enthusiasm for the discovery was powerfully self-reinforcing. The stock promptly doubled and rose by about tenfold over several years. It ultimately created severe losses for the short sellers and even contributed to the forced liquidation of one hedge fund. This experience underscored the types of impact (both positive and negative) that a new discovery could generate in the independent E&P sector. The large-acreage holdings in the Union Pacific Railroad land grant intersected this geologic trend, and as a result it would enjoy meaningful exposure to many of the dozen or so fields being discovered and developed there. For me, it sealed the case for going to Denver.

Just two months before undertaking that move, I visited Michel Halbouty in Houston, Texas, to obtain a firsthand perspective on the counterarguments to M. King Hubbert's more sobering view on the future potential of U.S., and ultimately global, oil and gas exploration and production. An article in the February 1980 issue of *Petroleum Independent Magazine* had prompted the effort to meet Mr. Halbouty. This meeting provided the basis for my First Boston research report published in May 1980 entitled *Talking to an Optimist—An Interview with Michel T. Halbouty on the Remaining U.S. Petroleum Reserve Base.*

Mike Halbouty passed away on November 6, 2004, at the age of ninety-five. Throughout his life, he was an imposing figure and petroleum industry leader. I used to say that if Hollywood were to make a movie about his extraordinary life, Dennis Weaver should cast in the title role, given the two men's uncanny resemblance to one another. By training, he was both a geologist and petroleum engineer. Michel Halbouty was a past president of the American Association of Petroleum Geologists and also received the DeGolyer Distinguished Service Medal of the Society of Petroleum Engineers. I had read two of his books, *Spindletop*, which recounted the discovery of the first giant oil field in Texas, and *The Last Boom*, a fascinating tale of the discovery of the East Texas field, which to date remains the largest conventional oil discovery in the lower forty-eight states (both co-authored with James A. Clark).

The core of his thesis during our discussion at his office in Houston was that stratigraphic traps still afford very significant potential for providing substantial new reserve additions in the United States. At the time, Halbouty used the term "subtle trap" to describe situations in which the encasement of petroleum has occurred not because of a distinctly identifiable structural subsurface feature but because the oil is trapped in a reservoir created by a broad regional unconformity in the earth's subsurface. He cited the real-world example of the East Texas field, with an approximate extent of forty-five miles in length and twelve miles in width. In this case, the trap of petroleum did not occur on a single large structure but instead was caused by a truncation or pinch-out of the formation on the eastern edge of the field and juxtaposition of the formation's western edge against impermeable rocks.

In the early 1980s, Halbouty frequently cautioned his petroleum industry colleagues against putting too much emphasis on pursuing what he called low-risk "corner shots" where the potential for reserve additions was too limited to address the pressing U.S. oil supply needs. In sum, Halbouty pointed out that since the petroleum industry's inception in the mid-1800s, total U.S. discoveries of oil in place amounted to approximately 450 billion barrels, of which about 110 billion had been produced to that point. He estimated remaining recoverable reserves at 40 billion barrels and added that he believed enhanced recovery technologies would ultimately enable recovery of another 100 billion barrels from the discoveries to date. He concluded by asserting that he was also optimistic that future discoveries of oil in place could once again at least equal all that had previously been found and evaluated up to that point. He further asserted similar optimism about natural gas. At the time, this degree of optimism was a highly contrarian view. It contrasted sharply with the broadening acceptance of Hubbert's peak oil (and gas) thesis.

Mr. Halbouty had an evangelical style of presentation that some found off-putting. Nonetheless, the substance of his message was too profound to dismiss. As I reread the points of his thesis of more than three decades ago, it is hard not to be struck by how time and subsequent human technical efforts have proved him essentially correct. As I describe later, the realization of his prediction is now actually occurring in a way that he undoubtedly did not fully anticipate. This is because there have been great advances in the petroleum industry's ability to map the earth's subsurface. There have also been breakthroughs in drilling and well completion technology to unlock oil from source rock where it is generated, as well as from tight rock adjacent to source rock. These developments along with much higher petroleum prices have multiplied by fivefold or more the economically viable resource potential that can be accessed. Consequently, Halbouty's fundamental optimism about the remaining U.S. oil potential is well on the way to being validated. Inspired at least in part by Halbouty's contagiously upbeat views, I published my report on the interview and prepared to move to Denver.

While actually making a trip through Denver in early June 1980, I saw in the airport a newspaper headline that read, "Exxon Sees 500,000 b/d from Western Slope Oil Shale by 1990." My reaction was that that gave me one more reason to make the move. On that particular point, I (and Exxon) could not have been more wrong. It was still the right decision but for other reasons. Within the next few years, Exxon and most of its peers had abandoned their Colorado oil shale projects because of the previously described costliness of maturing kerogen into useful synthetic oil. The town of Parachute at the center of oil shale activity began to resemble the early stages of becoming a ghost town. Its population of 1,300 just prior to Exxon's decision to pull out was more than halved in two years. Perhaps aptly named, that town needed a parachute to break the fall. As outlined later, this abandonment of oil shale projects by major oil companies would actually facilitate the opening up of a large new source of natural gas in the Piceance basin.

Merger-Mania I

Recognition develops that finding oil on Wall Street is cheaper than drilling for it.

As the early 1980s began to unfold, several trends were becoming evident that would contribute to a powerful reshaping of the U.S. oil industry and ultimately have knock-on effects on the structure of the international oil and gas sector as well. First, with the high oil prices prevailing in 1979–80, most petroleum companies plowed much of their expanded cash flows back into efforts to find and develop new oil and gas reserves in relatively more secure, non-OPEC areas. Many of these projects involved high-cost "frontier" environs, such as the deeper waters of the Gulf of Mexico, Arctic Alaska, the Gulf of Alaska, remote regions of Canada, and much deeper and more difficult formations in onshore basins of the United States. In many cases, these capital-intensive efforts were unsuccessful or only marginally successful. As First Boston's senior oil analyst, I was tracking these largely fruitless, or at least ultimately disappointing, conventional exploratory efforts year by year and basin by basin.

Of the fourteen largest companies that I was then analyzing at First Boston, only two were actually able to grow their reserves for the period 1978–81, and those involved significant acquisitions (Belridge for Shell Oil and General Crude Oil for Mobil). Without those transactions, Shell replaced only 86 percent of its reserves and Mobil 92 percent. For the remaining dozen companies, the operating records were even more dismal. Their average reserve replacement rate was reported to be just under 63 percent. In other words, for every barrel of oil being produced, less than two-thirds of a new barrel was being developed and added to these companies' reserves. In sum, it was a depressing and concerning time for the upstream U.S. oil and gas sector.

In the mid-1970s, I was a daily commuter through Manhattan's Grand Central Station, taking an escalator up to my office at 245 Park Avenue. One day it occurred to me that an escalator could be a useful metaphor for describing the nature of the "upstream" oil and gas exploration and production activities of the companies I was following. To pursue this conceptual analogy, I published a research report in March 1978 entitled *Up the Down*

Escalator—An Oil and Gas Exploration and Production Perspective. I opened the report with the following introduction from a book published in 1942 entitled *Oil Property Valuation* by Paul Paine: "The oil business is never static, and producing companies either grow or decline to the extent that they discover or acquire newly proved tracts." If drilled successfully, Paine explained, such lands provide new production growth, which is eventually followed by a decline in production and thus the need for a search for new properties to develop. Thus, in the absence of planned expansion and on ongoing efforts to maintain reserves, the producing company becomes a liquidating enterprise.

Because a mature, depleting oil field can typically produce between 5 percent and 10 percent (and occasionally more) of its recoverable reserves in a given year, at least the amount extracted needs to be found before a company holding a portfolio of such fields can show actual growth in reserves. Thus, for such an enterprise to grow is somewhat like trying to climb up the down escalator in a shopping mall (or a railroad station). To make progress up the escalator, one must advance faster than the rate of descent of the escalator. By the late 1970s, the rate of production declines had increased because of both reduced drilling activity stemming from the impact of oil price controls and low finding rates for new reserves even in existing U.S. oil fields. Thus, it was as if someone was turning up the speed of the down escalator just as the oil industry producers were trying to climb it. The situation was clearly suboptimal because fewer new petroleum reserves were being brought into production despite much higher expenditures for drilling and development.

Not surprisingly, finding and development costs for new reserves began to escalate very sharply. Between 1975 and 1981, there was a fourfold increase in finding costs per barrel for the major oil companies. This was caused by higher costs for every foot of well drilled as well as significantly fewer barrels of reserves added per foot drilled. Thus, the cost of staying in business and avoiding becoming a liquidating concern in the upstream petroleum sector was escalating rapidly.

At the same time, sharply higher oil prices had triggered enough destruction in the growth of demand for oil that institutional investors in public companies began to anticipate an inevitable decline in oil prices. Thus, many portfolio managers were not inclined to pay up in the stock market to fully discount the value of securities of oil companies, because of rising concerns about future oil price trends. Accordingly, the implied value per barrel of oil and gas reserves behind each share of public stock in oil companies was not appreciating commensurately with rising finding costs or with oil prices. In fact, in the early to mid 1980s in numerous cases, well-established oil companies were selling at about 50–60 percent of the net asset value that analysts were estimating for their reserves and other assets. Accordingly, over several years this valuation gap had widened to the point that it became compelling for financially strong companies to replace reserves by buying less financially well positioned companies holding such undervalued reserves in existing fields. This arbitrage gave rise to the expression that it was cheaper to find oil on Wall Street than in the oil patches of Texas, Louisiana, Alaska, and so forth.

In the summer of 1981, there began a wave of corporate merger and acquisition activity that over the next half decade would represent the largest corporate consolidation up to that point in the history of the petroleum sector. It began with a takeover of Conoco by DuPont. In a foreshadowing of trends to come, Conoco was actually put in play by a surprisingly strong response of its shareholders to a tender offer by the Canadian independent oil company Dome Petroleum. There was reason to be skeptical about the ability of Dome to actually consummate a full acquisition of Conoco, which was considerably larger. In fact, Dome's move may have simply been a ploy to see whether it could trade any acquired shares for Conoco's holdings in its Canadian affiliate, Hudson's Bay Oil and Gas. In any case, the speed with which the tender pool filled up and was then oversubscribed was a signal to the market that Conoco shareholders were ready to consider a more serious and credible bid.

Our First Boston client was DuPont, a chemical company looking to achieve backward integration on a strategic scale to meet its hydrocarbon feedstock needs. Stepping forward, it was successful in capturing Conoco, a prize that brought control of enough oil-equivalent energy production to equal (at least notionally) the company's entire petroleum feedstock requirements. This advisory engagement enabled First Boston's M&A team to demonstrate its ability to successfully help execute a technically complex, large, even historic, transaction. This was the first of a new wave of

THE COMBINATIONS OF COMPANIES THAT CREATED BIG OIL

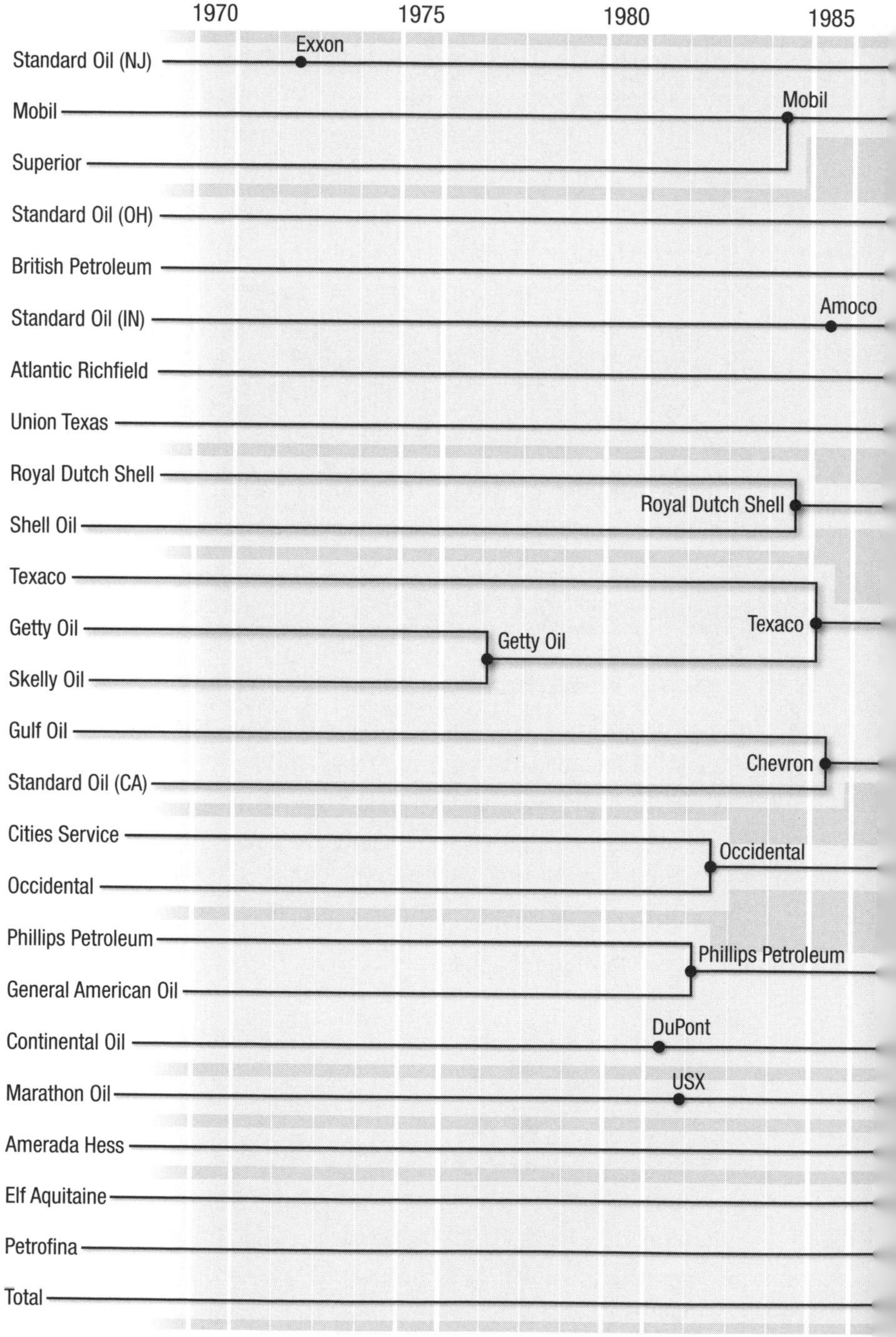

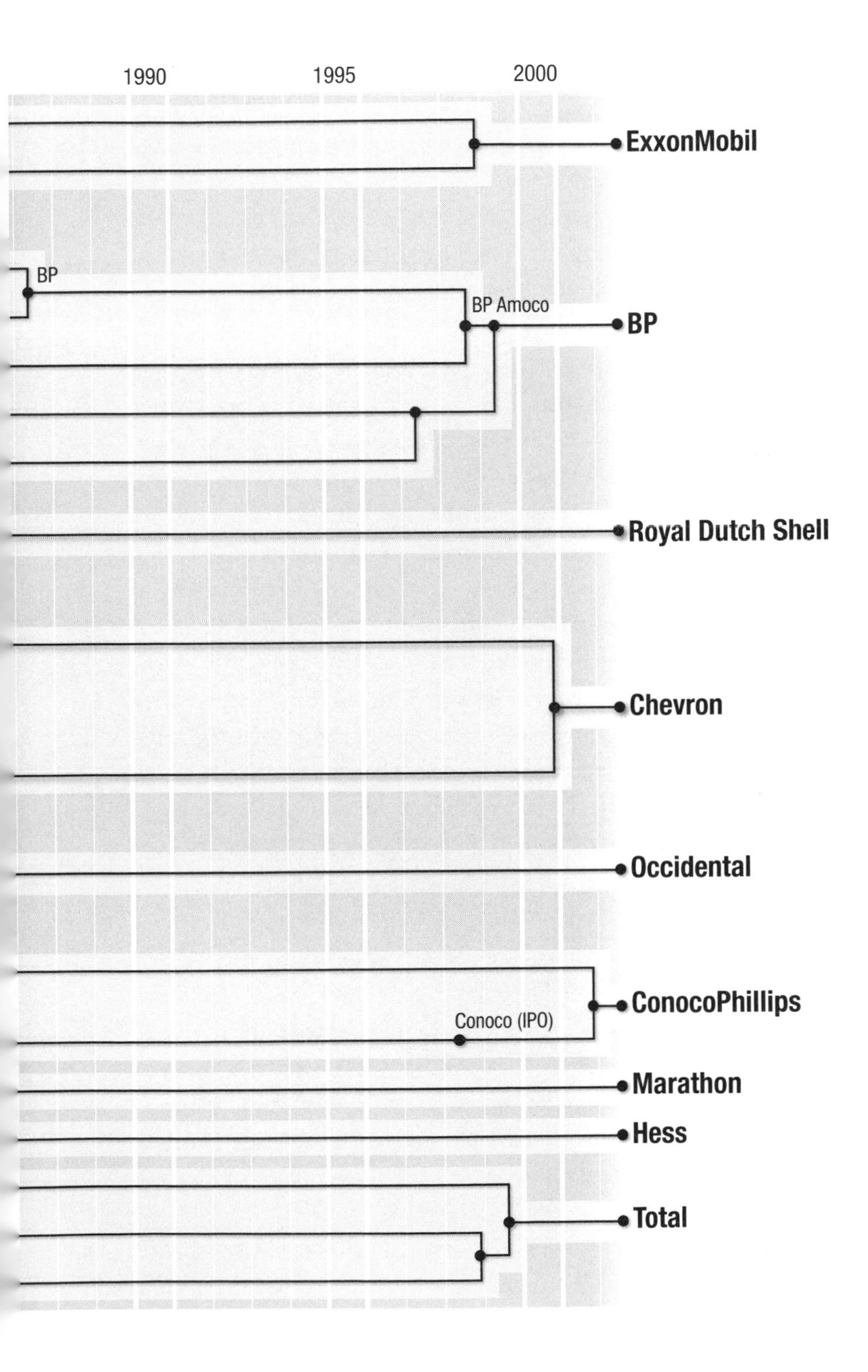
1990
1995
2000
ExxonMobil
BP
BP Amoco
BP
Royal Dutch Shell
Chevron
Occidental
ConocoPhillips
Conoco (IPO)
Marathon
Hess
Total

multibillion-dollar transactions in the integrated petroleum sector (that is, companies with oil production and downstream refining and petrochemical operations). Thus, it signaled to securities markets and investment bankers that previously presumed antitrust barriers to energy mergers were diminishing.

Not long thereafter, the acquisition of Conoco was followed by an entirely unsolicited and hostile bid by Mobil Oil for Marathon Oil (based in Findlay, Ohio) in late October 1981. Mobil's bid for Marathon significantly escalated both the tone and pace of Wall Street energy M&A activity. It immediately incentivized nearly all of the major investment banking firms to try to align with "white knight" candidates that might be able to provide a better alternative bid for Marathon Oil. This tactic was not going to be easily implemented, considering that Mobil was one of the four largest U.S. oil companies. However, the stakes were high, and the opportunity was not lost on Goldman Sachs, Morgan Stanley, and Lehman Brothers, among others.

The previous summer, Thomas Cassidy, First Boston's senior banker on the Marathon Oil (MRO) account, had organized a presentation for senior management of the company by Joe Perella, Bruce Wasserstein, Art Reichstetter, and me on takeover defense measures. A month or so later, we were formally hired as defense advisor and began developing an analysis of the company's assets, the largest of which was its Yates field holdings in the Permian basin of West Texas. Mobil's announcement of the offer was made on a Friday, October 30, and the Marathon board met the next day to consider its response. The proposal was a combination of cash and securities for Marathon's stock valued at $85 per share. This was a compellingly large premium to the previous market price. Thus, the most feasible option clearly was to find another buyer at a higher price. For the next several weeks, our First Boston team scoured the global universe of possible bidders. While there was obvious curiosity on the part of those we called, we were not finding meaningful traction with anyone we considered a credible alternative to Mobil. It seemed that other players in Mobil's peer group were still reluctant to take on the perceived risk of entering into a deal that might still involve lengthy antitrust litigation.

As we approached the deadline for expiration of Mobil's tender without a "white knight" acquirer, Bruce Wasserstein came up with an idea to buy time by approaching Ed Hennessey, the CEO of Allied Signal Corporation. Allied was small compared to Mobil, but it had a wholly owned oil subsidiary, Union Texas Petroleum (UTP), which Bruce felt could pose a creative alternative transaction to the Mobil offer and thus reset the tender clock under the then-prevailing SEC rules. With only a few days left for the Mobil tender to be triggered, Bruce, Joe, and I made a trip with Harold Hoopman and Elmer Graham, MRO's CEO and CFO, respectively, to New Jersey to meet with Allied Signal management. In addition to Hennessey, this included Clark Johnson, Union Texas Petroleum's CEO. The meeting was inconclusive at best, and on the late-night flight back to Findlay we were reluctantly adjusting our thinking to how to accept and hopefully somehow sweeten the terms of Mobil's offer.

We went to bed well after midnight, and around 5:00 a.m., I received a call from Elmer Graham telling me to get Bruce and rest of our team to the office by 7:00 a.m. In response to an earlier call by First Boston, David Roderick, CEO of U.S. Steel, was coming to discuss buying the company. By now we were down to less than two days remaining on the Mobil tender, so negotiations began in earnest and occupied most of Sunday. Around 3:00 p.m. that afternoon I received a call from Allied Signal's investment banker asking when Bruce would be available to address the Allied board meeting that had been called apparently after some kind of encouragement by Bruce following our rather inconclusive Saturday meeting. When I relayed this to Bruce, he said, "Tell him I'll call him tomorrow." This left our First Boston colleague, Brian Young, stranded at Allied Signal's corporate offices in New Jersey with the exceedingly awkward task of relating "what Bruce would say if he were here!"

In the meantime, our task of formulating a winning alternative turned on convincing U.S. Steel that it was worth their while to try to top Mobil's offer. This necessitated negotiating for a package of cash and securities that exceeded the value of Mobil's bid as well as providing for a consolation prize in the event that the Mobil tender proceeded. The latter involved providing U.S. Steel an option to purchase the company's Yates field outright on an asset transaction basis. This required us to determine a value to be paid for Yates that would stand the test of inevitable litigation regarding fair market valuation of the field. My prior research analysis on Yates along with an update that my colleague, Art Smith, had done recently provided the foundation for putting together this valuation. With it, the First Boston M&A team was able to complete a mark-to-market revised analysis in an incredibly short time.

My recollection of this phase of the deal is particularly memorable in that I was in a bad taxi accident on my way to brief the CEO of Marathon on our valuation of Yates. My discussion ended up being by telephone while lying on a table in the emergency room of a Chicago hospital having my bleeding arm sewn up with twenty or so stitches. Shortly thereafter at Christmastime, Harold Hoopman sent me a dark maroon tie with the Marathon logo on it. His accompanying note said, "If you don't like the color you can always use it as a tourniquet."

The Yates field "crown jewel" option proved decisive in countering the Mobil tender, but as expected, Mobil filed suit to have the court vacate the option. To make their case, Mobil's CEO, William Tavoulareas (now deceased), asserted in testimony to the court that we had underpriced the Yates field property to make it a bargain purchase for U.S. Steel. Also at the trial, the late Neil Armstrong, the astronaut who first walked on the moon and a Marathon board member, testified to his and the overall board's reliance on my expert advice regarding the valuation of the Yates field. My prior six years of detailed involvement with and analysis of this giant oil accumulation were brought to bear as I took the stand and was subsequently cross-examined for hours by Mobil's attorneys. When the judge quickly ruled in our favor, finding no evidence of a bargain transaction, U.S. Steel was able to consummate the merger. A few weeks later, in a decision that did not change the by-then-completed transaction, an appeals court reversed the lower court finding, though not because of anything wrong with the valuation. The appeals court simply deemed a crown jewel option illegal on its face. This ended crown jewel options "forever" (at least so far).

Marathon was actually the first of what would prove to be a series of such bids by a variety of corporate raiders. Over the next five or so years, about every six months there developed offers (often unsolicited) that in turn put into play General American Oil (acquired by Phillips Petroleum), Cities Service Company (acquired by Occidental Petroleum [OXY] following an abrogated contractual offer by Gulf Oil), Getty Oil (initially bid for by

Pennzoil, then acquired by Texaco), Superior Oil (acquired by Mobil), and Gulf Oil (acquired by Chevron). Finally, Phillips Petroleum and Unocal were both pressured by Mesa Petroleum into self-administered corporate reorganizations to unlock value for their shareholders. As a First Boston managing director and senior oil analyst, I along with my M&A colleagues had direct advisory involvement in more than three-quarters of these transactions.

As I have outlined earlier, the economic motivations behind most (though not all) of this activity centered on the idea that it had become compellingly cheaper to acquire oil reserves in the stock market than by drilling in many prospective hydrocarbon basins. T. Boone Pickens had become one of the early and high-profile advocates of this viewpoint. After commenting as an outside observer on the Mobil/Marathon hostile offer situation, Boone next surfaced as a primary initiator or catalyst successively in the sales, mergers, or forced restructurings of Cities Service, General American Oil, Gulf Oil, Phillips Petroleum, Unocal, and Diamond Shamrock Corporation. While Boone was not the prevailing acquirer in any of these situations, in each case Boone's company, Mesa Petroleum, emerged as a financial winner. Sometimes this resulted from gains on stock positions that Mesa had established prior to surfacing as a bidder. In other situations, the returns were further augmented by payments made for Mesa Petroleum to withdraw from the competitive landscape. The latter, known as greenmail, became a common component of transaction structures involving unsolicited offers and attacks on the poor or noncompetitive performance of certain corporate managements.

In sum, during the first half of the 1980s there was an extensive reconfiguration of the U.S. petroleum corporate sector. While he was by no means the only such activist focusing on the petroleum companies during this period, Boone Pickens was among the most adept at identifying vulnerable, undervalued enterprises, organizing a corporate raid, and articulating to other shareholders the merits of his ideas for change. By the mid-1980s, it had become evident that Mesa was unlikely to succeed as a hostile acquirer. Nevertheless, Pickens' imprint on the petroleum industry's restructuring was historic.

Several years later when Boone published his first autobiography, *Boone*, there was an interesting passage in which he gives an account of the conclusion of a confrontation between our First Boston client, Cities Service, and

Boone's company, Mesa Petroleum. This situation evolved in the spring of 1982 after Joe Perella discovered that Mesa was accumulating stock in Cities Service with an expected plan to launch a tender offer for the company in the near future. As defense advisors for Cities, Joe Perella and Bruce Wasserstein encouraged the CEO, Charles Waidelich, to call a meeting to brief his board. At that meeting, they convinced the board to authorize Cities management to launch its own preemptive tender for shares of Mesa Petroleum. Under the tender offer rules prevailing at that time, this would provide Cities an important timing advantage over any subsequent tender by Mesa. After Cities went public with its offer, I recall Boone saying it was a "popgun" and would have no real effect. In his initial response, Pickens provided a letter indicating his willingness to pay $50 per share for Cities Service. When it became clear that this was having no effect, he then proceeded to announce his own partial tender for 15 percent of Cities (all he could afford) at $45 per share. A few weeks later, over 48 percent of Mesa's shares had been tendered to Cities Service, and thus it was very close to exercising control of Boone's company in advance of Boone's ability to tender for Cities. With the Cities offer having a four-day advantage over that of Mesa, Boone was no longer calling it a "popgun."

However, after achieving important high ground in this M&A battle, the Cities board of directors called an executive session (that is, management was asked to step out of the meeting). The board then informed their advisors (First Boston and Lehman Brothers) that however nice it was that we had Mesa Petroleum somewhat on the defensive, the board had decided that they actually wanted to use this recently gained flexibility to find another buyer for Cities. After receiving this new guidance, we went to work and were able to negotiate a superior offer of $63 per share for the company from Gulf Oil.

The problem then became one of figuring out how to get Boone to step aside and pave the way for Gulf to acquire Cities Service. To do this, Bruce Wasserstein came up with the idea of having me take a red-eye flight to New York after planting the idea with one of Boone's advisors that Bruce and I would be meeting with the Gulf Oil CFO to recommend proceeding with both Gulf Oil's purchase of Cities and Cities' purchase of Mesa. In Boone's account, this had credibility since it would mean a bigger aggregate deal and thus payment of two fees for the advisors to Cities. He was told that it would be the first double takeover in history (Boone used the bird hunting term a "cluster shot"). The only problem was that we already knew Gulf had no interest at all in simultaneously doing the two deals. Thus, it was all a rather hollow bluff in an attempt to soften Boone up for the negotiation for Mesa to withdraw. Our concern was that if he knew Gulf's position, he might be inclined to hang around and continue to be an activist, something that was likely to cause real consternation for Gulf Oil.

In his book, Boone indicated that he only briefly had some concern that our implied plan for a bid for both Gulf Oil and Mesa could lead to the disappearance of Mesa into Gulf Oil. Shortly, a call between Boone and the Gulf Oil CEO, James Lee, assured him that buying both companies was not in the cards. In another account, it was reported that Mesa had also reached out to see whether Tom Brown, Inc., would launch a bid for Mesa to buy time to fend off Cities Service's bid. While that ploy did not materialize, Boone still was able to negotiate a settlement that included a $24 million profit for his trouble, along with his willingness to depart the scene. This he did the very next day with considerable media fanfare. For the advisors to Cities Service, it was a strong reminder that knowing a company board's ultimate goals is critical to render effective strategic advice. This was an early event in the evolution of American corporate governance toward boards of directors and shareholders holding management much more accountable for operating performance and taking appropriate action when management falls short. Boone Pickens' agitating actions became a primary catalyst that advanced that mindset.

The story of the sale of Cities Service involved one final chapter. In early August 1982, apparently because of internal concerns of its management that it had agreed to pay too high a price, Gulf Oil unilaterally abrogated the merger contract, weakly citing notional concerns about antitrust issues regarding downstream operations. Accordingly, as advisors to Cities Service, we were forced back to the market to find another buyer for the company. The alternatives were now few, and Armand Hammer, Occidental Petroleum's CEO, sensing this, successfully moved with a proposed merger utilizing various forms of new OXY securities. The transaction value was about 20 percent lower than the level of Gulf Oil's bid. Decades later, Chevron, the subsequent acquirer of Gulf Oil, had to pay almost one billion dollars in damages (most ironically to Occidental) after losing the litigation over Gulf's abrogation of the 1982 merger contract.

Another one of the companies that Boone Pickens was able to put in play was General American Oil, a midsize independent producer based in Dallas, Texas. Having published research reports in 1978, 1981, and early 1982 analyzing the company's assets and positioning, I had come to know William Barnes, General American's CEO. When Boone decided to try to buy the company, he hired Morgan Stanley as his advisor. This proved fortuitous in that it eliminated the only other investment banking firm that Bill knew on Wall Street. Upon arriving in New York City late on the Sunday night of December 12, 1982, I had a message to call Bill "regardless of how late it might be." Upon reaching him about midnight, I was informed on the spot that he wanted to hire First Boston as his takeover defense advisor; I immediately conveyed the news to Joe Perella and Bruce Wasserstein.

The General American Oil deal picked up where the Marathon Oil and Cities Service transactions had left off. In a matter of relatively few days, we solicited "white knight" candidates to top Boone's offer. In short order, Phillips Petroleum emerged with a significantly higher bid, but it came with a surprising and unusual condition. Phillips wanted General American to ensure that Boone would go away "happy," meaning that he should receive an extra payment in return for providing a written assurance that in the future he would not interfere with General American "or its affiliates." While the de facto greenmail payment was a provision executed as part of the final sale to Phillips, the anti-activist protection sought by Phillips proved only transitory. It was ultimately legally unenforceable about three years later when Mesa Petroleum launched a raid on General American's "affiliate" Phillips Petroleum. Pickens' move forced a restructuring of that enterprise to unlock greater shareholder value.

Another transaction predicated on the idea that oil reserves could be captured more economically via Wall Street than by drilling involved Texaco's acquisition of Getty Oil. In fact, over a career of more than sixty years, J. Paul Getty had pioneered the concept of achieving growth in oil reserves per share by consistently using periods of low stock prices to repurchase publicly held shares in his companies. Through these open market purchases of outstanding stock, he was able to boost reserves as measured by barrels per share. This was an outgrowth of his early success in gaining greater control of Pacific Western and Tidewater Oil, respectively, during and after the Great Depression. It was a technique he later used with Getty Oil as well as Skelly Oil.

As an analyst back in October 1975, I had interviewed Mr. Getty at his home in Sutton Place in Surrey, just outside London. It was about nine months before he died. The North Sea oil boom was in full swing, and Getty Oil was participating with Occidental Petroleum in developing the giant Piper oil field discovery. Paul Leibman and I met for about two hours with Mr. Getty in his library where he was receiving daily reports on the North Sea drilling. The "pay phone booth" for which frugal Getty was renowned as having installed for the use of his houseguests was just down the hallway. Having read his autobiography, I was well aware of Getty's philosophy about acquiring reserves via oil stock purchases. Accordingly, I asked him why he had not moved to acquire all of the shares of Skelly Oil, since he already owned a large block of the company. He responded that he believed that "two dogs going through the forest could scare up more game than one." He seemed to imply that his holding in Skelly was, at least for the time being, an exception to his overall petroleum stock acquisition philosophy. After the announcement of his death the following June, I strongly suspected that Getty management might be more likely to buy the rest of Skelly Oil as a way to improve their reserve replacement performance. A year or so later that, in fact, happened.

By 1983, with major oil company merger and acquisition activity on the rise, Getty Oil's management was increasingly at odds with the Getty family interests led by Gordon Getty, one of J. Paul's sons. A preliminary announcement of a deal for Pennzoil to acquire Getty Oil was stirring up interest by other potential buyers. Texaco hired First Boston to assist in making a higher offer. My presentation to the Texaco board as it considered authorizing the bid focused on the strategic merits of Getty's holdings. These consisted of the giant oil accumulations in the Kern River field in California, the Piper field in the U.K. North Sea, and the Neutral Zone concession between Saudi Arabia and Kuwait. Texaco was successful in topping the Pennzoil proposal. Curiously, this high-profile transaction drew almost instantaneous criticism as evidence of the excesses of capitalism by the Soviet press agency TASS. Interestingly, in the later postcommunist and emergence of state capitalism era, the Russian oil company Lukoil purchased the downstream petroleum product marketing assets, including the branding rights to Getty Oil. In a dramatic reversal of fate, Pennzoil's shrewd litigation and maneuvering obtained a favorable "hometown" legal venue with a Texas jury trial. This culminated in a highly controversial finding of tortuous interference and a $10 billion judgment against Texaco. While this assessment was ultimately settled for about one-third of the court's finding of damages, it involved a painful bankruptcy process to facilitate the negotiations. The lesson for other oil company CEOs was to never underestimate the ability of a team of creative strike suit attorneys to engineer a miscarriage of justice. Subsequent reforms in the Texas judicial system have helped address some of the excesses that confronted Texaco in this case.

The M&A Wave Creates a New Set of Opportunities

Upstream asset rationalization of ownership escalates.

At a remarkably early point in the petroleum corporate merger cycle, my colleague Jim Parkman developed a critical insight that would form an important part of the business of Petrie Parkman in the 1990s and beyond. Jim recognized that after the wave of large corporate mergers there would be a growing need to rationalize the ownership of petroleum holdings at the asset level. This would involve producing oil and gas fields, undeveloped prospective acreage, natural gas plants, and other aboveground equipment to support production as well as oil- and gas-gathering pipeline systems and hydrocarbon storage facilities. In part, Jim saw this rationalization process as a natural follow-on to corporate combinations where there would be strong motivations to monetize assets deemed "noncore" to these larger newly merged enterprises. In addition, the highly fragmented nature of the ownership of U.S. reserves, production, and passive minerals meant that a growing number of private and corporate parties would inevitably be wanting to liquefy their holdings as their financial needs and goals changed over time. Not long thereafter, Tom Edelman, another former First Boston investment banking colleague, underscored the same theme to me with the statement, "Petroleum property ownership in the U.S. is incredibly fragmented; it screams for rationalization." Tom Edelman left First Boston in 1980 and has ever since pursued a highly successful entrepreneurial career focused on that screaming opportunity.

Jim Parkman, soon after he developed his insight about rationalizing the ownership of petroleum assets, approached me to help him convince First Boston about an idea that the firm should organize a "petroleum property divestiture practice," which he could lead and manage with just a few supporting bankers. First Boston embraced the idea, and over a five-year period beginning in 1983, Jim and his team successfully executed several dozen transactions with an aggregate value of over $5 billion. These individual deals typically ranged between $25 million and $200 million, and they were distributed over a wide variety of oil- and gas-producing basins across North America. Another great insight of Parkman's was that the divestiture practice would provide useful intelligence about the regional and geological differences in evaluating corporate portfolios of oil and gas reserves as an input to the structuring of the much larger corporate mergers. By being continuously in the market for individual assets, our First Boston team had the ability to appreciate better than many of our competitors the valuation impact of changes in external market conditions, regulatory rules, and improvements in production and well completion technology.

One of the relatively early opportunities to pursue this asset monetization advisory strategy involved working with Brownlie, Wallace, Armstrong and Bander, the BWAB partners, on a transaction to realize a significant portion of the discovery value of their holding in the Anschutz Ranch East discovery in the Overthrust Belt. Rather unusually, the transaction was formulated as the sale of a stream of production payments to an income fund–oriented oil and gas operator. The latter company and other income partnerships like it went on to become active buyers of oil and gas properties in many of our producing asset sales transactions throughout the balance of the 1980s. This transaction marked the beginning of a relationship with the BWAB founders that in subsequent years led to other opportunities to work with Jim Wallace and Ray Brownlie.

Chapter 4

TURNING POINTS

The Oilman's Road Trip—Russia and China, September–October 1983

Given flawed economic incentives, "crisis amid plenty" is both possible and even likely.

During a twenty-day trip in the late summer and early fall of 1983, I made a trip that completely circled the globe. On this occasion, I had the opportunity to accompany a dozen or so independent oilmen from the Independent Petroleum Association of America (IPAA) on a U.S./Soviet Union/Peoples Republic of China–sponsored exchange trip touring those countries. Among others, the group consisted of present, former, and future IPAA chairmen, including John Miller, A. V. Jones, Gene Ames, and Paul Hilliard. Both countries we visited were communist regimes with a deep commitment to central planning to achieve national economic priorities. However, the contrast between the Soviet Union and China was one of night and day. Russia and its satellite nations under an ailing President Andropov were a scary and depressing remnant of the superpower adversary about which I had come to know something at the height of the Cold War while serving with the U.S. Army in Germany in the late 1960s.

When the Soviet Union invaded Czechoslovakia on the night of 20–21 August 1968, I was awakened in my quarters in Darmstadt just south of Frankfurt about 4:30 a.m. that morning with a phone call to mobilize with my battalion to the field with the attention-getting warning that "this is not an exercise." In a matter of hours, we were headed east on the autobahn toward the German/Czech border to set up communications channels for U.S. Army Europe. Shadowing us along the way were Russian "liaison officers" seeking to monitor our frequencies and routes of deployment. Under an agreement between NATO and the Soviet Union, these individuals were based in West Germany with the objective of ensuring against an "accidental" commencement of war. They were supposed to be prohibited from leaving the autobahn and other major prescribed routes. Nevertheless, they routinely ventured throughout the West German countryside pursing their secondary mission as Soviet forward observers (spying) on U.S. Army Europe activities and operations. Against this backdrop, one can better appreciate some of my later reservations regarding the Soviet Union.

By the early 1980s, the Soviet economy was showing almost no real growth and was laboring under the burden of central planners who were essentially oblivious to, and even disdainful of, economic market realities. Throughout our trip, we witnessed firsthand grocery stores bereft of bread, milk, and other basic foodstuffs, as well as widespread evidence on the city streets of dysfunctional alcoholics. The state-run security apparatus was clearly designed to coerce and propagandize an increasingly impoverished citizenry.

Our main visit to oil facilities was in Baku, Azerbaijan, where our hotel accommodations were highly reminiscent of sparse quarters I had known some fourteen years earlier as a military advisor in Vietnam. When we deplaned in Baku, we walked through a reception gauntlet of far-less-than-friendly people that strongly reminded me of a similarly strained reception in the Philippines on my way into Vietnam. I commented on this to our Russian guide, who also acknowledged a degree of discomfort. Even more revealing was the shockingly decrepit state of production facilities along the shore of the Caspian Sea. However, many of the Azeri people themselves, upon realizing we were Americans, were warm, outgoing, and interested in providing us an enjoyable and informative visit. Most interestingly, it featured a tour of a memorial to Azeri leaders who had been executed by the Russian communist leadership in 1917. There was also a fascinating trip to a restored trading outpost along the ancient "Silk Road" connecting China to Europe.

Our final day back in Moscow is one I will remember for the rest of my life. It began with a visit to the Soviet Oil Ministry. We had a briefing on Soviet petroleum development plans that lasted about an hour. Shortly before our trip, the Russian leadership had announced plans to increase oil production from 11 million barrels per day to 15 million barrels per day by the decade's end. In 1983, this was big news in that the Soviet Union's production already had overtaken that of Saudi Arabia, and here was the government promising to widen the gap even further!

In preparation for our trip, I had researched the sources of Soviet production and had come across an engineering paper that indicated the Samotlor field was the largest Russian field. In fact, it was a contender to be the largest producing field in the world at that time. The field was producing over 3.5 million barrels per day, thus accounting for about one-third of total Soviet output. However, the problem was that evidence had emerged that Samotlor was beginning to produce increasing quantities of water along with oil. I surmised that increasing water production in such a giant field was most probably a precursor to the onset of an irreversible production decline. Accordingly, I challenged the Soviet Oil Ministry presenter's forecast of such large increases in overall Soviet output. In fact, I raised the question early in the meeting, again in the middle of the meeting, and persisted once again at the end. In each case, the response was "not to worry, it is all under control." Finally, with some exasperation, they further explained that they really knew how to handle the water!

Years later it emerged that the Soviet idea of controlling water was to install submersible electric pumps to handle the rising fluid levels as water cuts increased. For the very short term, this enabled field workers to meet the arbitrary production quotas handed down from Moscow. However, the problem with this ill-conceived solution was that in fairly short order the pumps literally began to suck the water preferentially to the well bores. Thus, it effectively bypassed the oil, creating stranded (and thus unproducible) oil accumulations throughout the field, in what amounted to a gutting of the field's natural producing mechanism.

Over the next decade, Samotlor's decline rates accelerated, and ultimately the field output dropped by more than 90 percent to a mere 250,000 barrels per day. This is a classic case study that demonstrates how arbitrary top-down production quotas issued by a central planning regime and lacking any provision for a strong bottom-up technical validation (that is, a "sanity check") had seriously impaired economic performance. It was a real-life example of the old adage, "You get what you incentivize people to do." The problem of handling water at the Samotlor field was not an isolated case of a central planning agency generating arbitrary and counterproductive quotas that in turn generated a number of economic misallocations. Numerous other examples of distortions are detailed in an excellent analysis, *Crisis amid Plenty: The Politics of Soviet Energy under Brezhnev and Gorbachev,* by Thane Gustafson. Subsequently, Michael Economides, a Houston-based economist, has also detailed some of the same issues as well as other more recent ones in *From Soviet to Putin and Back: The Dominance of Energy in Today's Russia.* Past may not always be prologue, but certainly in the case of Russia, knowing something of history is helpful.

The earlier-referenced 1983 discussion with oil ministry officials regarding Soviet production prospects was sufficiently intense that several of the U.S. independent oil producers attending the meeting still remind me of that debate I had each time our paths cross, usually at oil industry conferences. Following the oil industry meeting, our entire group of IPAA representatives went to the U.S. Embassy in Moscow for a debriefing on our experiences with the desk officer of the day. For about an hour and a half, we related our impressions from meetings, tours, and other inputs gained over the nine days spent throughout the country. Given revelations in recent years about the past Soviet bugging of the U.S. Embassy, it is interesting to contemplate how what we discussed and especially that which is described below might have been received by those Russians whom we now know were capable of listening in.

By way of conclusion to our debriefing session, the American Embassy desk officer offered the following summary, which I paraphrase: "Let me share with you what I think is now going on here in Soviet Russia. A while ago, a dissident poet and entertainment celebrity died. There was no announcement of his death and no announcement of his funeral. At the memorial service a hundred thousand people showed up to pay their respects." He paused for the effect of that statement to sink in and then stated, "This outpouring of high regard for this poet drove the authorities crazy!" He then added, "For further perspective, I'll share with you the four lines of poetry for which he is well known and even loved. It is roughly translated as follows:

> Let us hope they never discover life on Mars
> Because if they do
> The Soviet Union will be
> Poor forever.

The desk officer went on to observe what both the poet and his many Russian mourners deeply appreciated about those four lines: "No matter how remote the threat, the Soviet Union would always defend itself against such perceived distant enemies even to the impoverishment of its people in terms of guns versus butter." This seems likely to have been Vladimir S. Vysotsky, a Russian singer, songwriter, poet, and actor. He became widely known for his lyrics, which featured social and political commentary in often-humorous street jargon. Even now he exerts an influence on Russian entertainers, who seek to emulate his styles and influence.

From the U.S. Embassy we boarded our bus to the airport. As we took off for our eleven-hour flight to Beijing, our group broke into spontaneous applause. Sitting next to me, Jude Wanniski (now deceased), then the back-page editor of the *Wall Street Journal*, leaned across the aisle to explain to a group of puzzled Chinese women volleyball players that we were pleased to be "getting out of Soviet jail" and heading to China. When we arrived in Beijing, we were taken to the Chinese State Guest House, where Margaret Thatcher had been visiting the prior week to negotiate the eventual return of Hong Kong to China. Thus began our experience of observing the dramatic contrast between the communist economic systems of these two very different countries and societies. At dinner that night, one of our group with a shortwave radio informed us that while we were in the air, a Russian fighter plane had shot down Korean Airlines Flight 007, killing hundreds of innocent passengers. The Russians claimed that the plane had wandered into Soviet airspace. On hearing this, we were immediately reminded of our U.S. Embassy desk officer's statement only twenty-four hours earlier that however remote the threat, it would be countered strenuously regardless of the cost to Russian citizens. Given this persistent "guns versus butter" bias, the Soviet Union's reputation as well as the economic lot of its citizens suffered mightily once again from this horrific incident. A little over a week later, our group flew home to the United States over essentially the same flight path and in the same type of plane as the Korean airliner. Having by then read the accounts of the plane's violent plunge into the ocean, we felt a direct connection to the tragedy.

The Soviet/China trip is probably best characterized as a study of sharply contrasting systems, because after our arrival in Beijing, much of what we saw and experienced was upbeat and genuinely encouraging. To be sure, there was still evidence of China's challenges and problems stemming from decades of economic misallocations, which were due to the communist penchant for central planning just as in Russia. However, one could not help but be deeply impressed by the energy, dedication, work ethic, and sense of purpose of the Chinese people. Furthermore, the first signs of a new attitude and approach under the leadership of Deng Xiaoping were already becoming evident. Our itinerary included visits to many of the usual tourist attractions, including the Forbidden City, the archival site of the terracotta warriors, and a trip to part of the Great Wall. A most impressive aspect was a tour of Beijing Museum, where we witnessed many examples of how advanced Chinese society was over many prior eras dating back several thousand years. The evidence of the creativity of Chinese people was manyfold, with water pumps and other clever equipment designs dating back two thousand and more years, as well as noteworthy artistic and architectural accomplishments.

With regard to the petroleum sector, the timing of the trip coincided with a renewed initiative by the Chinese government to open up to foreign companies access to offshore acreage in the East China Sea. This was an early part of Deng Xiaoping's program of economic liberalization. In retrospect, it is evident that the optimism about the hydrocarbon prospectivity of acreage offshore China's extensive coastline was more excessive than warranted (or at least it was premature, given the then-existing state of exploration technology). Subsequent drilling to evaluate the blocks of acreage granted to foreign oil companies resulted in a string of generally discouraging or condemning dry holes. Nevertheless, there is no doubt in my mind that this was an earnest desire to undertake economic development with a more modern and liberalized, market-based approach.

At the time, China was producing about two million barrels per day, largely from its giant Daqing oil field, but its economy was only consuming about one million barrels per day, with the balance being exported to other Asian markets to earn China much-needed foreign exchange credits. I also believe that the Chinese leadership realized that the Daqing oil field was mature and would soon be entering an irreversible natural decline. That very predictable prospect and the looming need to develop new energy sources as China liberalized its economy were motivating factors for changes. Throughout our nine-day visit to China, our Western petroleum group encountered a more open-minded and constructive reception. It was a striking contrast to our experience at the Soviet Oil Ministry in Moscow.

Another memory that anecdotally underscores the differences between our Soviet Russian and Chinese experiences involved meeting a fifteen-year-old high school student in Shanghai. The first night of our arrival in that city, Jude Wanniski and I were taking a walk around to familiarize ourselves with the area near our hotel. We were approached by this young man who inquired whether he could walk along with us to practice his English. Since this in turn gave us the chance to learn some things about the city, we were happy to oblige. During the course of the evening, the young man proved to be so engaging, informative, and interesting that he was invited to join us, and he effectively became our informal guide for the three days we were in the Shanghai area. In fact, our group's regard for him grew to the point that a few years later when he finished high school, several of our party put together a fund to support his attendance at a U.S. college, where he subsequently excelled and graduated.

In closing this section, I will briefly jump ahead in time to relate something about my lasting impressions of the Russian trip experiences. I returned home from that tour deeply impressed by the size of the country, the humanity of its people, and its extensive oil and gas resource potential. However, I also was convinced that the evolution of the Soviet Russian petroleum industry away from a centrally planned communist economy into one more compatible and competitive with Western oil enterprises could require multiple decades to accomplish. This remained my mindset a half decade later in 1989 when I was in the process of forming Petrie Parkman & Co. with my First Boston colleague Jim Parkman. The Berlin Wall had come down, and market recognition was dawning that the end of the Soviet Union as we had known it was looming. I received a call from a former First Boston client, Joseph Gruss, an elderly, highly successful private investor who had originally emigrated from Poland in the late 1930s. Mr. Gruss asked me to meet him in New York to discuss an investment idea. Even in the 1970s and 1980s, he still spoke with a heavy accent that often challenged my ability to understand him, but I always listened attentively because I knew he was a wise man.

Joseph Gruss' story could be the basis of a fascinating novel. He was Jewish and born in Lvov (which is now called Lviv) in the Ukraine. He attended business school in Poland. By the 1920s he was in business in Warsaw, where he witnessed the severely adverse knock-on effects of the German hyperinflation and its aftermath, including the empowerment of Adolf Hitler. Reading the tea leaves leading up to the Kristallnacht era of Jewish suppression, he was able to use his ownership of a travel agency to obtain exit visas for his wife and himself to come to America in the late 1930s. However, it was at the cost of abandoning much of his business affairs as well as financial assets in Poland. In the United States, an old family friend was able to secure a desk for him at Wertheim & Co., where he began trading securities. Over the ensuing decades, he was very successful in rebuilding his wealth. Among other projects, he partnered with the legendary Andre Mayer of Lazard to form Toreador Royalty out of the mineral interests of the Matador Land and Cattle Company. The latter had originally been formed with Scottish investors in the 1880s by the famous cattleman Charles Goodnight. That company held petroleum prospective lands across the North Texas Panhandle. Gruss also accumulated significant stock positions in Skelly Oil and Getty Oil that were ultimately worth many multiples of his investment when those companies were merged and sold in 1977 and 1983, respectively.

In our meeting in April 1989, Mr. Gruss proposed that together we should set up an oil company in the "new Russia" that he felt would "inevitably" emerge from the then-prevailing chaos. He offered to finance the venture if I would move to Moscow and run it. He wanted me to negotiate to obtain access to as many electrical logs as possible of dry holes drilled by the Soviets and then hire a Schlumberger log engineer to determine where the Soviets had somehow missed (or bypassed) economic oil. He said that with the Iron

Curtain having come down "everything will change, and we can make a fortune." He was eighty-nine at the time but had lost none of his entrepreneurial focus and insight. Mr. Gruss was absolutely correct in his vision; fortunes have indeed been made, but as a dreadful series of experiences of William Browder with his Hermitage Fund illustrates (as well as that of numerous others), the realized rewards have often been, at best, elusive for many Western capitalists schooled in businesses accustomed to operating under the rule of law. At the time, I did not particularly see the challenge of Russian oligarchs and the emergence of "rule by Putin decree," but my late 1960s experiences in the U.S. Army in Germany and on the 1983 Soviet Union oilman's trip cautioned me to stay focused on North America, where a degree of rule of law still applied. Although I decided to decline Mr. Gruss' business offer, he ranks to this day high among the mentors I have been fortunate to know. He taught me much about having a global vision of forces at work that can cause energy investments to become successful.

Oil Price Collapse, January–July 1986

It is often darkest before the dawn.

In July 1980, William M. Brown and Herman Kahn of the Hudson Institute published an exceptionally prescient article in *Fortune* magazine entitled "Why OPEC Is Vulnerable." Brown was the institute's director of energy studies, and Kahn, its founder, was highly regarded for his ability to think outside the box about macro drivers of global economic trends. At the time, while OPEC was widely viewed as bordering on being omnipotent, these contrarian thinkers cogently made the case that "the organization, never as powerful as it looked, may crumble in the 1980s." Their analysis was based on their view of longer-term supply/demand trends and likely intergovernmental relations of the OPEC members and was very much at odds with the prevailing consensus. Other commentators were much more sanguine, and another five or six years would pass before the Hudson Institute's thesis was fully validated. While OPEC's demise was not total, it came perilously close to slipping away as a functional organization. In the final analysis, the Brown-Kahn assessment is a stunning case study, well worthy of the attention of today's prognosticators on any matters involving the relationships between economic drivers and aggregate human behavior within organizations comprising diverse countries and cultures.

The half decade or so following the second oil price spike of the late 1970s to early 1980s involved trends that were initially somewhat similar to those that followed the first. High oil prices induced cumulative economic drag that ultimately curtailed oil demand. OPEC tried somewhat unsuccessfully to mitigate the downward pressure on oil prices by reducing production. Their efforts were thwarted by a pattern of cheating as individual OPEC members tried to game the system by "sneaking" above-quota volumes of oil to market. Not surprisingly, as depicted in the frontispiece, oil prices were drifting lower by 1983. Furthermore, U.S. independent oil producers, oil service companies, and energy investors, having "seen something like this movie" in the mid to late 1970s, adopted the motto "Stay alive 'til '85 or Chapter 11 in '87!" It worked out somewhat differently than that phrase suggested, but as is true of all humor (especially the gallows variety) it did touch on the underlying insight of this increasingly discouraging situation.

As was the case for the realities of my oil tanker experience back in 1973–74, by mid-1985, the fundamentals of oil supply in Saudi Arabia dictated a new course of strategic moves for that country away from trying to reduce its exports to sustain an artificially high oil price. At its peak in the early 1980s, Saudi Arabia's oil output was over 10 million barrels per day, but in the ensuing half decade it had borne the burden of the vast majority of OPEC's quota cuts. By the summer of 1985, its output fell below 3 million barrels per day, and oil prices were still under pressure because of economic sluggishness in the oil-importing economies of the world. With output at about 2.5 million barrels per day in July of that year, there were reports that Saudi Arabia was even struggling to produce the oil it needed to generate enough associated natural gas to run the country's electric power generating systems. Saudi Arabia in the summer without air conditioning is (like Houston, Texas) a very uncomfortable place.

Something had to give, and it did. In early August, King Fahd announced that Saudi Arabia would no longer incur virtually the sole burden of oil production cutbacks. Its OPEC partners were served notice that cheating on agreed-upon export quotas would no longer be tolerated. They would either have to step up and bear their proportionate share of the required production cutbacks to balance supply with demand or Saudi Arabia would take steps to recapture its historic share of the oil market. This was a major warning shot across the bow aimed at much of the rest of the OPEC membership. It signaled that an oil price war and a painful market share realignment were imminent unless most of the other cartel members demonstrated uncharacteristic oil quota discipline. The latter was not going to transpire easily. For other OPEC members, more than a decade-long practice of promising one level of production and delivering quite another was deeply ingrained. Furthermore, the disparity between the pressing financial needs of the more populous oil-exporting countries versus those with relatively low populations and high oil output reinforced the inclination to ignore the Saudi warning (actually, it was a virtual demand) for OPEC-wide sharing of the production quota reduction. Thus, oil prices began to decline in late 1985 to the low $20s per barrel range. However, they did not actually spiral downward until the first half of 1986, as the players tried to sort out the extent of the demand problem and how serious was King Fahd about regaining some of the country's historical market share. At year-end, that question was still unanswered, but market observers were appropriately very uneasy.

As 1986 opened, oil prices began slipping into the sub-$20 per barrel range, and it became clear that Saudi Arabia was indeed determined to regain its share of the market. Its annual revenues from oil exports had declined from over $100 billion to a projected level of less than $20 billion, well below what the country needed to fund its social programs and economic development. Through the winter and spring months, the downward pressure was relentless, and by early summer, oil prices were hovering in the $11–13 per barrel range. The impact on oil stocks was predictably negative. In April, an independent exploration and production company that we at First Boston were advising took bids for its sale after a 75 percent decline in its public market value. By July, with the oil price then below $10 per barrel, the other members of OPEC were ready to hear Saudi Arabia's proposal to restabilize the oil markets. OPEC met in late July 1986 and shortly afterward announced a new agreement to curtail production with the goal of restoring prices to an $18–20 per barrel range. However, the many preceding years of broken agreements and disregarded quotas gave rise to widespread doubts in consuming countries that individual OPEC members possessed the requisite discipline to deliver on promised quota compliance.

In late July, I received a call from Alan Abelson, the editor of *Barron's*, asking what I thought of the OPEC announcement. Recalling my West Point military history courses, which had taught me that often the tide of battle could turn for the better just when it seemed most discouraging and least likely, I observed that I disagreed with the then-consensus on the likelihood of ongoing OPEC weakness. Picking up on my "it is often darkest before the dawn" analogy, Alan (who is now deceased) suggested doing a cover story interview for the upcoming week. Thus began over a decade-long challenging but thoroughly enjoyable and enlightening dialogue with him and *Barron's* on petroleum supply/demand as well as a variety of other energy issues. The case for oil price recovery that I outlined in that *Barron's* interview was predicated on my conviction that the Saudis had inflicted so much punishment on fellow OPEC members that strong compliance with the agreement was actually quite likely for at least the next year or so. I also believed that during the same period, oil prices would probably average lower than in recent years and thus would still be a net stimulative to restoring growth in the global economy and a resumed expansion in oil demand.

Finally, I believed that oil prices in the first half of 1986 had dropped much lower than the Saudis had ever intended. There were indications that the oil minister, Sheik Yamani, had assumed oil markets would restabilize when oil dropped to around $15 per barrel, a level at which he expected much of the North Sea production to be shut-in because he had been informed that the break-even profitability price for North Sea oil was about $15 per barrel. While that level was indicative of a break-even level on a reported book accounting basis for some North Sea oil fields, the cash flow break-even level that would actually trigger actions to shut down North Sea production was actually less than $5 per barrel for many North Sea fields! The latter price was a level that Saudi Arabia did not want and could not even afford for very long. Accordingly, it was my view that the prospects were better than most market observers were predicting for a workable cooperation among OPEC members for the intermediate term. That said, I must admit I still experienced anxious weeks in September and October while slack seasonal demand continued to test OPEC's resolve. By year-end, however, my "it is darkest before the dawn" hunch was validated, and OPEC had returned to an operating mode supportive of higher prices and reduced volatility.

Stock Market Crash, October 1987

Doubling up at a market bottom works (as long as it is a real bottom)!

When compared to the most recent financial crisis of 2008 involving capital market declines, the stock market crash in the fall of 1987 now seems relatively mild, even merely episodic. However, at the time, it was feared to be the onset of something quite ominous, and it initially bore enough resemblance to the early phases of the stock market crash of 1929 to leave many observers deeply concerned.

My own recollections of the lead-up to the October 19, 1987, debacle are branded into my memory by efforts we were making at First Boston just before then as the lead manager of the initial public offering (IPO) for Union Texas Petroleum (UTP). As a consequence of a partial (50 percent) leveraged buyout of UTP by Kohlberg Kravis & Roberts (KKR) in the fall of 1985, joint control of the company was shared by the previous 100 percent owner, Allied Signal Corporation, and some of the funds of KKR, of which the principals were Henry Kravis and George Roberts. The late 1985 timing of their original investment could not have been much worse, preceding as it did the culmination of the 1986 oil price collapse. However, to the great credit of UTP's management, led by CEO Clark Johnson, the company did adapt to and survive the challenges of that difficult year. By early 1987, UTP was actually in a position to contemplate going public. Unfortunately, a variety of factors and issues delayed our ability to bring the IPO to market in the spring of 1987, and in those days, launching a major equity offering during the vacation-oriented summer months was not advisable.

Accordingly, we reset our plan for a fall execution of the IPO but did so with trepidation that the market might not be as accommodating as it could have been earlier in the year. The reasons for this concern were many. First, we would be coming to market just after the first anniversary of the bottoming of oil prices in 1986—with memories of those volatile days accordingly refreshed by retrospective media accounts. Second, currency markets had been exhibiting rising concerns since April and were becoming increasingly viewed as threatening to the U.S. economic recovery. Third, other equity market placements were already encountering resistance and pricing sensitivity. Just two weeks earlier we had found the First Boston–led pricing of a convertible preferred stock for Triton Energy to be exceptionally challenging. Afterward, I warned the CEO, Bill Lee, not to count on being able to come back to market for more funding for a considerable period of time (in my mind at least two years).

Despite these issues and concerns, it was clear that both Allied Signal and KKR had strong motivation to move forward with a program to liquefy their UTP holdings. UTP had an international operating footprint with U.K. North Sea oil production and Indonesian liquefied natural gas facilities, as well as oil and gas production in Pakistan. Thus, we decided to reach out to both European and Asian markets with a road show starting before Labor Day, followed by a trip circling the United States that would culminate in a late September pricing of the IPO. The European swing started slowly in part because continued rising currency concerns and other securities market weakness were distracting investors as we wound our way through Edinburgh, London, Paris, Frankfurt, Zurich, and Geneva. It did not improve much in Asia as a second management team visited Singapore, Hong Kong, and Tokyo. By the time we were back in the United States after Labor Day, we knew we were dealing with an uphill climb to build a book of buy orders for the IPO. Other developments were underscoring the prospect of a weakening market outlook. For example, a few days prior to the IPO pricing, I received a call from a client who wanted to know whether he should sell his one-million-plus share position in another independent oil company. Based on my perspective, I responded that if he wanted to sell anytime over the next year or two, he should do it very soon given the looming overall market concerns. He sold later that day at a price that proved to be the best to be had for his stock for the next two-plus years.

Meanwhile, our progress on the UTP deal was still languishing. The proposed pricing was originally set at $18–21 per share. At the time of our SEC filing, it was reduced to $16–18 per share because of deteriorating overall market sentiment. At this level, the midpoint of the deal still contemplated over a $300 million capital placement, a goal that was fast fading.

On the morning of the day that we proposed to price the offering, our orders in the book amounted to only about 40 percent of the deal. This meant we needed to find more than another $200 million of additional buyers to have any chance of pricing that day. Late order builds for IPOs are not unusual, but achieving one of this magnitude would be extraordinary for that era. We went to work, and by that afternoon we had barely enough demand to propose a downsized deal to the current shareholders. However, to have any chance of success in the aftermarket, we also recommended to Allied Signal and KKR that they not sell any secondary shares themselves and that the price should be $14 per share, a further discount to the already reduced price range of $16–18 per share. After considerable debate, hand-wringing, and some pointed client criticism of our efforts, the deal was priced as recommended on September 29. Later that night, I received a call from Clark Johnson, who thanked me for our tenacity in getting the IPO launched. He, better than anyone, understood that extraordinary efforts had been expended to build the book of buyers in an incredibly difficult market. The professional relationship that I forged with Clark during that transaction ultimately led to other opportunities to work together when Jim Parkman and I formed our own firm a few years later.

The UTP IPO closed a few days later, and almost two weeks afterward the October 19 stock market crash occurred. The weekend prior to that fateful Black Monday, I was at West Point for my twentieth class reunion. That event's weekend highlight was watching the "walking in space films" of two of my classmates, Mike Mullane and Woody Spring, who had become NASA astronauts. Because the prior Friday stock market close had been extremely unsettling, I decided to spend Monday at First Boston's office at Park Avenue Plaza in midtown Manhattan. Watching the rising panic of normally disciplined and well-composed human beings as selling decisions became totally divorced from rational analysis is an experience I will never forget. That night I joined two friends, Tony Pace, a hedge fund manager, and Luis Mendez, head of trading for First Boston, for dinner to discuss implications of the stock market collapse. Another friend, my former boss, Leo Mandrakos, also joined us. He was running a portfolio at the time and had spent the afternoon assessing the damages. We debated well into the night whether this Black Monday would prove to be the precursor of another depression akin to 1929. Such was not the case, but at that moment it was far from certain. To wrap up our deliberations, we each scrawled our predictions on a napkin about where the market would be a year out. No one embraced the super bear case.

Not surprisingly, UTP stock fully participated in the stock market collapse and in a matter of weeks was trading below $7 per share. At my year-end performance review, Jim Freeman, First Boston's director of equity research, observed something along the lines that given the aftermarket performance of UTP, my career as an oil analyst was probably close to being over. I recall responding that he might be right, but that he should know about all the investors having bought UTP on the offering that I had also persuaded to double up their positions below $8 per share. About a year later, UTP was back at $14 per share, and those who had added to their positions realized a worthwhile profit. In turn, I learned a most valuable lesson regarding longer-term investing: doubling up following a sharp decline can work as long as one can afford to be patient and ultimately prove to be right about the fundamental factors at work.

Another special relationship that developed during this period involved William I. Lee, founder and CEO of Triton Energy. As in the case of Union Texas, Triton would later become an early client of Petrie Parkman. I had first met Bill in 1984 when Triton discovered two oil fields in the Paris basin. The Villeperdue field was the larger of the two, at about 100 million barrels, and the Chaunois field was perhaps one-third of that. For a small independent producer such as Triton, these reserve additions were substantial, and I realized that the stock had the potential to appreciate significantly in value. Accordingly, I took several trips to France to better understand the geology of the play and extent of Triton's holding. As it turned out, the visit to the Chaunois field was particularly interesting in that it underlay an extraordinarily gorgeous country villa that was originally built by Louis XIV's finance minister. I was told that when the king saw what his underling had built (apparently with the king's funds), he had him arrested and put in jail. He then directed that there be built for the king an estate as beautiful as this one but that it should be even bigger. Hence, legend has it, Chateau de Fontainebleu was built.

Upon returning from the Paris basin, I published an equity research report recommending Triton Energy as an interesting speculative buy. Well after the report was circulated to First Boston's institutional clients, Brett Haire (a trader on the firm's desk) decided to buy a block of Triton warrants. These were rights to buy the stock that had been issued as a sweetener to a high-yield bond offering that had been done by another investment bank a year or so before the Paris basin discoveries. Because the warrants provided the ability to buy at a fixed price, they inherently provided a leveraged exposure to Triton's appreciation potential. As the company continued to drill up its French discoveries, the stock essentially doubled in price and the warrants appreciated about fourfold. In the end, they were sold back to investors in France, who somewhat belatedly became enamored with the Paris basin play.

Bill Lee was later described by *Forbes* magazine as "Lucky Bill Lee," and in a sense he deserved the title because in large part, as my mother is fond of saying, "people make their own luck." Over the course of his storied career, Bill had demonstrated repeatedly that he had a nose for finding oil and gas. He was involved in the discovery of the two large gas fields in Australia, two more valuable oil fields in France, two significant oil and condensate accumulations in Colombia, and ultimately two giant gas/condensate plays in Equatorial Guinea in West Africa and in the offshore neutral zone between Thailand and Malaysia.

I sometimes reflect on how surprised Bill's father would be by all Bill accomplished, as he lived life to its fullest. On a trip to West Point in the late 1990s, Bill shared a reminiscence with me about an earlier visit he had made as a teenager to that part of the world. In the late 1930s, his father put him on a train in Texas headed to the Peekskill Military Academy in New York on the east bank of the Hudson River. His father's parting admonition was something to the effect, "Son, when you get there, I want you to stand on the river's edge. Look across the river and you will see the U.S. Military Academy at West Point, where there are many high achievers. Look to the right, and upriver you will see the prison known as 'Sing Sing,' where people are sent to contemplate the error of their ways. I want you to know that I am not quite sure where you will end up!"

In recent decades, Bill would probably have astonished his late father. In recognition of his pioneering oil exploration efforts in France, Bill received the Legion of Honor, France's highest award for a non-French citizen. In my assessment, he deserved even more credit than he was ever accorded for building Triton into what ultimately proved to be a compelling acquisition for Amerada Hess Corporation (now Hess).

Finally, following Bill's investment decisions underscored for me the value of seeing the second-order implications of evolving oil industry trends and developments. For example, as a heavy user of seismic data to define exploratory prospects with state-of-the-art geophysical tools, especially in areas of difficult and challenging terrain, Bill came to appreciate new advances that were occurring in the equipment to gather such data. This led him to invest in a company called Input/Output. That company was a leader in developing much-improved, lightweight helicopter-transportable equipment. These advances made much higher quality geophysical assessment of rugged terrain possible and less costly, in many cases for the first time. He made this investment at a difficult stage of the petroleum price cycle when no one else would step up. Then it also was a contributing factor to a period of financial stress for Triton in the early 1990s, which led to an advisory role for Petrie Parkman. Ultimately, however, Input/Output was a meaningful success and a credit to Bill's contrarian intuition as well as investment acumen.

Breaking Away

A nonmarket–share driven platform creates focus, operating flexibility, and increased competitiveness.

By 1987 it was becoming clear that the oil and gas markets for M&A transactions were moving away from the megamerger consolidation wave involving large capitalization companies that had dominated the first half of the decade. At least for the moment, most of the logical combinations for the largest oil companies had been consummated. M&A advisories had, by then, largely morphed into corporate reorganizations, such as the Boone Pickens–initiated restructurings of Phillips Petroleum, Unocal, and then Diamond Shamrock. Furthermore, the arbitrage opportunities involving undervaluation of petroleum securities had diminished considerably versus the earlier years of the decade, partly because of SEC-mandated improvements in reserve disclosures and widening recognition by investors of the intrinsic value of oil reserves in a growing world economy. At First Boston, a significant change occurred when Bruce Wasserstein and Joe Perella decided to leave the firm and form their own investment banking boutique. Shortly thereafter, they invited Jim Parkman and me to interview to join them, and we both briefly considered the possibility. However, as we subsequently compared notes on this proposition, Jim and I each came to the conclusion that staying a while longer at First Boston would be better. We realized that there would be increased opportunities within our current roles given Bruce's and Joe's departure, but we also expected that there might be opportune changes at First Boston.

In the late 1970s, First Boston's CEO, George Shinn, had negotiated to buy White Weld's London-based investment banking operation, which was jointly owned with Credit Suisse. As part of this transaction, Credit Suisse held a significant minority equity interest in First Boston. We suspected that ultimately Credit Suisse would either move to acquire by merger the rest of First Boston or put its minority stake up for sale to another party that would look to acquire all of First Boston. In the fall of 1988, Credit Suisse and First Boston announced an agreement to do a full combination, and it was completed in early 1989. The weekend of the public announcement, Jim Parkman and I were on a flight to Germany to meet with a client in the petrochemical sector who was considering an acquisition strategy for U.S. oil and gas assets. We quickly concluded that the Credit Suisse move was probably the event for which we had been waiting.

In February 1989, Jim and I resigned from First Boston and formed Petrie Parkman & Co. Our vision was to create a boutique investment bank focused on providing advisory and transaction execution services solely in the energy sector. I planned to open an office in Denver, Colorado, with institutional investment research and capital market capabilities. Jim was ready to relocate from New York to Houston and open an office focused on providing advice on mergers, acquisitions, and divestitures. We were committed to not being market share driven but instead wanted to address the needs of fewer clients with a clear strategic focus. We positioned ourselves as having a Denver-to-Houston axis through the U.S. oil and gas "patch" with an ability to be especially responsive to the needs of clients in both of those cities. These locations accentuated the firm's ability to be readily available and quickly responsive to clients in Dallas and Midland, Texas; New Orleans and Lafayette, Louisiana; Oklahoma City and Tulsa, Oklahoma; Los Angeles, California; and prospectively Calgary in Canada.

During the first week of March, Jim and I visited Clark Johnson at Union Texas Petroleum. A few months earlier in November 1988, Clark had astounded me by pulling me aside after an analysts dinner meeting to say that he wanted to be the first client when I formed my own firm. At that point, the First Boston combination with Credit Suisse had been announced, but neither Jim nor I had given any indication to anyone of our thoughts about forming a boutique. Clark was as good as his word. In early March 1989 when Jim and I met with him to outline our reasons why he should hire us, he abruptly stopped us and presented his draft of a proposed engagement letter. It contemplated a broad strategic advisory on both capital markets and upstream asset matters.

Later that year, we arranged to take Union Texas to Tokyo as part of a Petrie Parkman–organized M&A symposium. It provided a useful follow-up to the company's IPO road show of two years earlier. An amusing incident in that conference speaks volumes about the cultural differences between American and Japanese views of M&A activities. Just as I was about to give my overview presentation on the state of energy M&A in North America, my translator cautioned me that because I would be using certain words that had no equivalent in the Japanese language, he would need to be creative in conveying some of what I had to say. I asked him to give me an example of this dilemma. He then pointed out that at the time (late 1980s), there was no concept for unsolicited or hostile corporate mergers and acquisitions in the management-by-consensus business culture of Japan. Accordingly, when I would refer to corporate M&A transactions he would be using the Japanese word for airplane hijackings to convey the concept of a change of corporate control!

Petrie Parkman's strategy and business model were built around our two core skill sets. For eight consecutive years, I had been ranked as the leading exploration and production analyst in the annual survey of *Institutional*

Investor Magazine. That investment research–based profile provided us with a platform upon which to build a position in selected capital markets activities. Executing on this idea became even more feasible when Paul Leibman and Stuart Wagner joined us a few months later. Jim Parkman's previous success in building the oil and gas asset divestiture practice at First Boston gave us confidence that we could quickly establish a profitable business helping companies monetize reserves and other assets that no longer were core to their operations. Given all the changes that had occurred over the prior decade, we knew that there was a significant backlog of properties fitting into this category.

This view was validated in short order when Louisiana Land & Exploration requested of First Boston that Jim Parkman remain fully involved in an ongoing sale of Rocky Mountain and Gulf Coast assets. It was shortly thereafter that I received a call from Jim Ukropina, CEO of Pacific Enterprises, asking us to handle a divestiture of assets that the company had previously acquired from Sabine Royalty Co.

This briefly presented us with a capacity problem, since at that time we had only three employees: myself, Jim Parkman, and my longtime, remarkably versatile assistant, Janeen Hogan, who was busily setting up our office space, books, and records and addressing the myriad of issues associated with becoming an operating entity. Jim was fully engaged in executing the Louisiana Land divestiture transaction. The problem began to be resolved when, later the same day as the Pacific Enterprises call, Randy King, a highly experienced senior petroleum engineer with Netherland Sewell Associates, called Jim to indicate that he had decided to join us. Because Randy was based in Dallas, we opened a temporary office for him to focus on the Pacific Enterprises divestiture. Now that we had a pipeline with a variety of client assignments, it was evident that we needed additional professionals. That requirement was fulfilled when Jon Hughes and Nick Gardner joined us soon thereafter as key founding colleagues of the firm.

With our first three client engagements, Petrie Parkman was generating positive cash flow within our first three months and was profitable in our first year, even after Jim and I satisfied an expensive noncompetition obligation stemming from our early departure following the Credit Suisse/First Boston merger. At year-end, we sent deal mementoes to Clark Johnson at

Union Texas Petroleum, Leighton Steward and Rick Bachmann at Louisiana Land & Exploration, and Jim Ukropina at Pacific Enterprises in grateful acknowledgment of their critical votes of confidence and the key roles that these engagements had played in the successful launch of the firm.

In retrospect, 1989 was a propitious time to form an investment banking boutique focused on the energy sector. There were only a few other entities operating along the lines of focused services that we were contemplating, and none had our particular configuration. On numerous occasions in those early days, there were skeptics who asked us, "Do you really think you can make a go of it with a single industry focus?" Our consistent reply was yes because the energy sector is so large and is such a key element to almost all other aspects of economic activity. Furthermore, we believed that the sector's complexities would create numerous examples of valuation disparities and market inefficiencies that called for resolution or arbitrage through well-conceived and organized transactions.

Another factor that gave us confidence occurred when Bruce Wasserstein offered very early on to fund our launch with a $5 million capital injection for a 49 percent interest in the firm. We decided to decline this generous offer, in large part because of the organizational and regulatory technicalities of effectively affiliating with Wasserstein Perella. However, it did motivate us to explore another option for funding our new firm. Accordingly, I approached Robert Day, a founder and the CEO of Trust Company of the West, with the idea of selling him 20 percent of our company based on the valuation implied by Bruce's offer. While it was not so obvious at the time, Robert paid us a huge favor by replying that he felt that new enterprises such as ours would be best funded by its founders. Taking his advice, Jim and I then resolved to provide the founding equity capital, and I advanced a loan to cover our initial working capital needs. For the first couple of years, we both lived on our personal savings, but we had the benefit of full and flexible control of the enterprise without the constraints that would have inevitably been attached to having otherwise uninvolved, third-party investors. Had we pursued the alternative route, the evolution of Petrie Parkman would have been altogether different and inherently more limited.

Chapter 5

GAME CHANGERS I

Kuwait Invasion/First Gulf War, 1990–1991

The consequences of perceived improbable, or black swan, events are often far greater than many expect.

In May 1990 at a meeting of Arab summit leaders, Saddam Hussein delivered a very threatening speech to the attendees, expressing his anger at Kuwait's and Saudi Arabia's high level of oil production versus their official OPEC quotas, which were crowding out Iraq's ability to move its oil to market at an acceptable price. Saddam's sensitivity on the issue was heightened by the debts he had incurred in the Iran-Iraq war. He indicated that Iraq had borne a disproportionate burden among the Sunni Arab nations in prosecuting that eight-year war with Shia Iran. Although Kuwait and Saudi Arabia had helped finance his military campaign, he was deeply resentful that they had not forgiven their loans to Iraq and had incurred none of the casualties in a conflict that bore an unsettling resemblance to the human agony of World War I. His neighboring countries were now increasing oil production well above their quotas in a manner that both kept oil prices depressed and limited Iraq's market access to sell its oil. Furthermore, he indicated that they should address the production (and possibly debt forgiveness) issues quickly or suffer the consequences. While the content and acrimonious nature of the meeting were reported in the *Petroleum Intelligence Weekly* a month or so thereafter, the oil and securities markets did not seem to reflect any heightened or unusual concerns about these threats. As May passed into June, commodity and major stock markets were generally stable, even relatively sanguine.

During the first week of July, I was on a three-person panel at an energy conference organized by the University of Houston. The other panel members were Georgetown University professor Madeleine Albright (subsequently, U.S. Secretary of State) as moderator and the CEO of a mid-sized oil field services company. The latter entity was based in Houston but had an international marketing footprint that included several Middle East countries. The format for our panel discussion called for opening statements by the moderator and each of us as presenters, to be followed by a question-and-answer (Q&A) period. I was asked to set the stage by providing a perspective on the global oil supply/demand outlook and its implications for oil prices for the balance of 1990 looking into 1991.

I proceeded to outline an assessment of new volumes coming to market netted against natural declines in established fields. It all pointed to no big change in the fundamental outlook unless some new factor disturbed the equilibrium that the market was then discounting. I wrapped up my comments with one cautionary observation: if Saddam Hussein were to follow up on his warnings from the May Arab summit meeting, it could and probably would represent the type of exogenous event that would upset the rather fragile supply/demand balance I had just outlined. I provided no statement of probability of this scenario occurring and simply identified it as a risk factor worth monitoring because in the event that Saddam were to make good on his ultimatum, it could have a significant impact on oil pricing prospects.

When the oil field services CEO got up to speak, he almost immediately took issue with my characterization of Saddam Hussein as a potential threat to global oil supply balances. He indicated that his company had a long history of operating in the Middle East, including in Iraq. He added that he personally knew Saddam and did not believe that there was any risk at all of Iraq punishing its neighbors Kuwait and Saudi Arabia over the issues I had identified. I was somewhat taken aback that what I considered a fairly low-key caveat to an otherwise mainstream oil price forecast elicited such a strong rebuttal. In effect, it became almost the core of his presentation. I noticed that Madeleine Albright also seemed surprised and was not quite sure how much emphasis to put on our difference of opinion in the Q&A period. During that session, I took the opportunity to reiterate my points regarding what had been reported regarding Saddam's assertions and to point out that I was not making a prediction. I was simply observing what the consequences might be if what I considered a low-probability event were to occur. My main point was that the nature of the threat had been reported in only one relatively specialized petroleum trade newsletter. Accordingly, the threat had probably not been broadly considered and thus not well discounted by commodity and especially securities market decision makers. Accordingly, I saw it as a possible scenario with the potential to have a meaningful impact on markets if it were to become a reality.

Almost exactly a month later, on August 2, I was vacationing on the coast of Maine standing on a marina dock about to board *Calliope*, a newly built lobster boat that was returning from sea trials. As the boat pulled alongside the slip, someone said to me, "Did you hear what happened last night?" I replied, "No, what?" The person told me, "Iraq invaded Kuwait last night and has overrun the country in a matter of hours. The emir of Kuwait has fled across the desert to Saudi Arabia." I turned to the dock attendant beside me and said, "Fill the boat up, that's the last cheap fuel we'll see for a while," and indeed that proved to be the case.

When I returned to my office in Denver the following week, I encountered another example of the second-order effects of this unprecedented geopolitical event. Vintage Petroleum, a privately held independent oil company cofounded in 1983 by Charles C. Stephenson, Jr., had filed an initial public offering (IPO) in early 1990. The road show to sell the deal had launched in the spring but had not attained the requisite purchase orders to execute a successful underwriting. Thus, from mid-June through July, the deal was relegated to the back burner awaiting better market conditions. Iraq's invasion provided just the needed tailwind to lift energy stock prices, in general, as well as institutional receptivity to Vintage's new issue story, in particular.

In retrospect, it is evident that Iraq's invasion of Kuwait is an example of the type of an exogenous (or black swan) event that can have powerful positive (and at other times negative) implications for energy sector participants. The specific probability factor associated with such an event is typically low to the point of being de minimis. However, the likelihood of a somewhat comparable event occurring somewhere globally in a given time period is actually not insignificant. Given this opening, Charlie Stephenson successfully executed his IPO, one of the very first of a new era for oil and gas initial public offerings. Over the following decade and a half, he went on to build Vintage Petroleum into a multibillion-dollar independent oil producer with operations in California, Texas, and Argentina that was ultimately acquired by Occidental Petroleum in 2006.

For many of my West Point classmates still serving in the U.S. Army, the first Gulf War was their final active engagement as they were completing their military careers. Seven of them were brigade commanders in the remarkably successful campaign to expel the Iraqi Army from Kuwait. Some seven years later, when I mentioned this to the foreign minister of Saudi Arabia, with whom I was meeting in February 1999, he remarked, "That was America's finest hour; you came, you accomplished the necessary task, and with it finished, you went home quickly."

Return of the IPO Cycle: Rebuilding the Publicly Traded E&P Universe

The regenerative powers of incentivized intellectual capital should not be underestimated.

In 1979 as an oil analyst looking to stake out coverage of a relatively new petroleum investment category, I identified sixty-two institutionally suitable, publicly traded independent oil and gas producing companies focused on the upstream sector of the United States. Within a year or two, it became clear that this was to be a high-water mark for domestic E&P players. By 1981, a process of attrition had become evident that would translate into over a two-thirds shrinkage in this group of companies, ultimately reducing the total to only nineteen freestanding companies by early 1987. That year marked the low-water line. That fall, the initial public offering of Union Texas Petroleum marked the beginning of an upswing in issuance of equity for new enterprises in the E&P sector. Over the next decade, this expansion wave would take the universe of publicly traded independents back through the old high to a new one totaling some seventy-four distinct entities by 1996.

As the Vintage IPO experience demonstrated, the process was not always smooth or easily predictable. Not surprisingly, the October 19, 1987, stock market collapse closed the IPO and new equity financing windows for several quarters. In fact, the planned spinoff of Enron Oil & Gas (now known as EOG Resources) from its yet-to-be ill-fated parent was initially scheduled to price on "Black Monday," October 19, 1987, but was not actually executed until about a year later. The next notable launch of a new E&P enterprise was that of Vintage Petroleum, which, as already recounted, struggled until it found an open market window because of the oil price lift in the immediate aftermath of Iraq's August 1990 invasion of Kuwait. That benefit to oil securities markets proved fairly temporary in that an oil price decline following the quick and decisive expulsion of the Iraqi Army from Kuwait dashed investor expectations for upstream oil and gas economics for another year or so.

Nevertheless, by early to mid 1992, forces had begun to align to favor the formation and flotation of a series of new E&P enterprises. First, while oil prices remained well below the pre–Gulf War highs, they began to show signs of greater stability at a level around $20 per barrel. Second, rising optimism began to develop that an eventual end to the seemingly ever-elongating natural gas bubble was coming into view as the surplus in natural gas supply was being worked off. Third, it was becoming clear that a group of entrepreneurial leaders in the petroleum sector had survived and were emerging from the corporate consolidation of the 1980s with a renewed vision about exploiting onshore, lower-48 petroleum resources. Many of these properties had been neglected, abandoned, or divested (through sales down the corporate food chain) by the larger companies to independents as the major oil companies redirected their exploration and development efforts toward larger reserve targets internationally. The combination of these three factors set the stage for a very powerful IPO underwriting cycle in the E&P sector for the next half decade. This proved to be a significant source of business for Petrie Parkman, given our investment research focus on the exploration and production sector and our collective in-depth knowledge of North American geologic basins.

The launch of Basin Exploration in 1992 validated the idea of equity market receptivity to the more favorable fundamentals that were unfolding for independent oil and gas producers. Basin had been formed in the late 1980s by Michael Smith, who had dropped out of the chemical engineering program at Colorado State University to pursue real estate ventures. Before long, that decision led him to recognize that the subsurface mineral–based real estate of the Denver-Julesburg basin east of Greeley, Colorado, might well provide a superior set of investment exposures. As occurred with many

SHRINKING (AND THEN EXPANDING) UNIVERSE OF LARGE INDEPENDENT COMPANIES

Number of independent E&P companies with market cap over $150 million

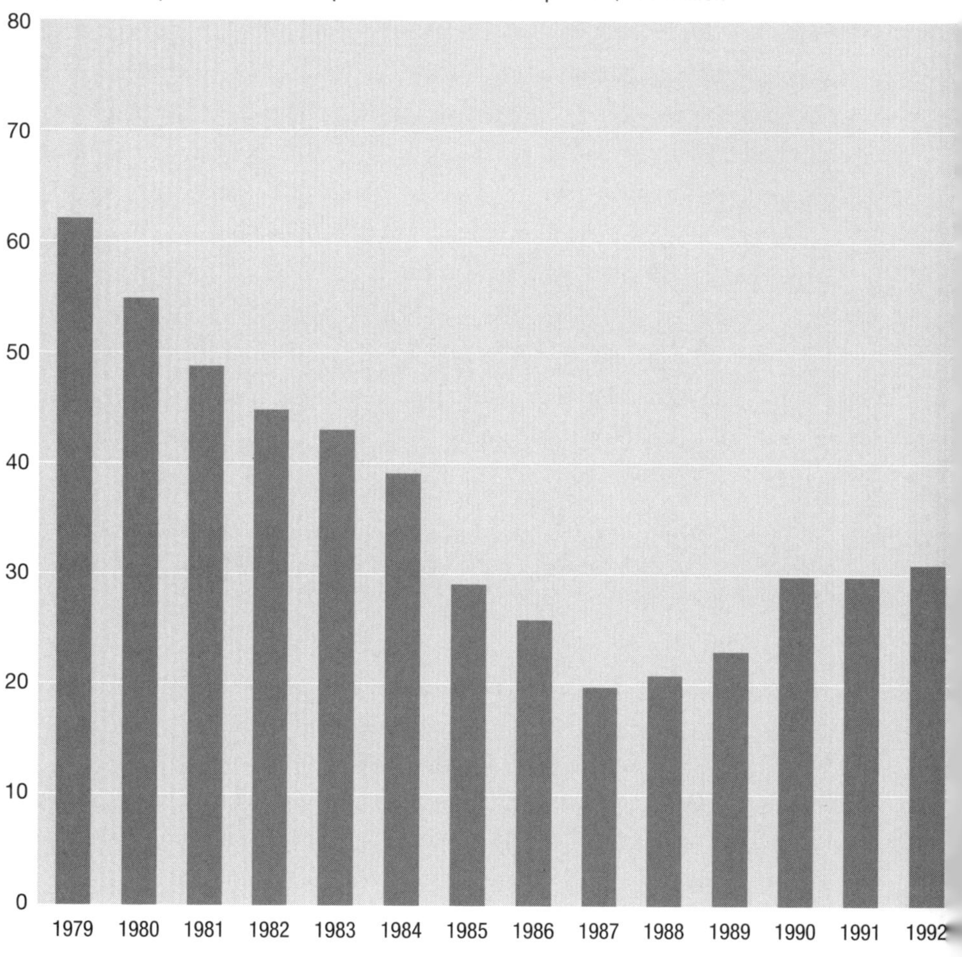

SOURCES: Research by First Boston Corp. and Petrie Parkman & Co.

upstart independents, Michael experienced his fair share of setbacks and challenges, but by 1991, he had come to realize that having a publicly traded platform would provide much-desired flexibility in financing his energy ventures. About that time, Michael began a series of discussions with me and others regarding what it would take for Basin Exploration to become a competitively positioned public company. Completion of this IPO in the spring of 1992 was quickly followed that fall by the launch of HS Resources, another

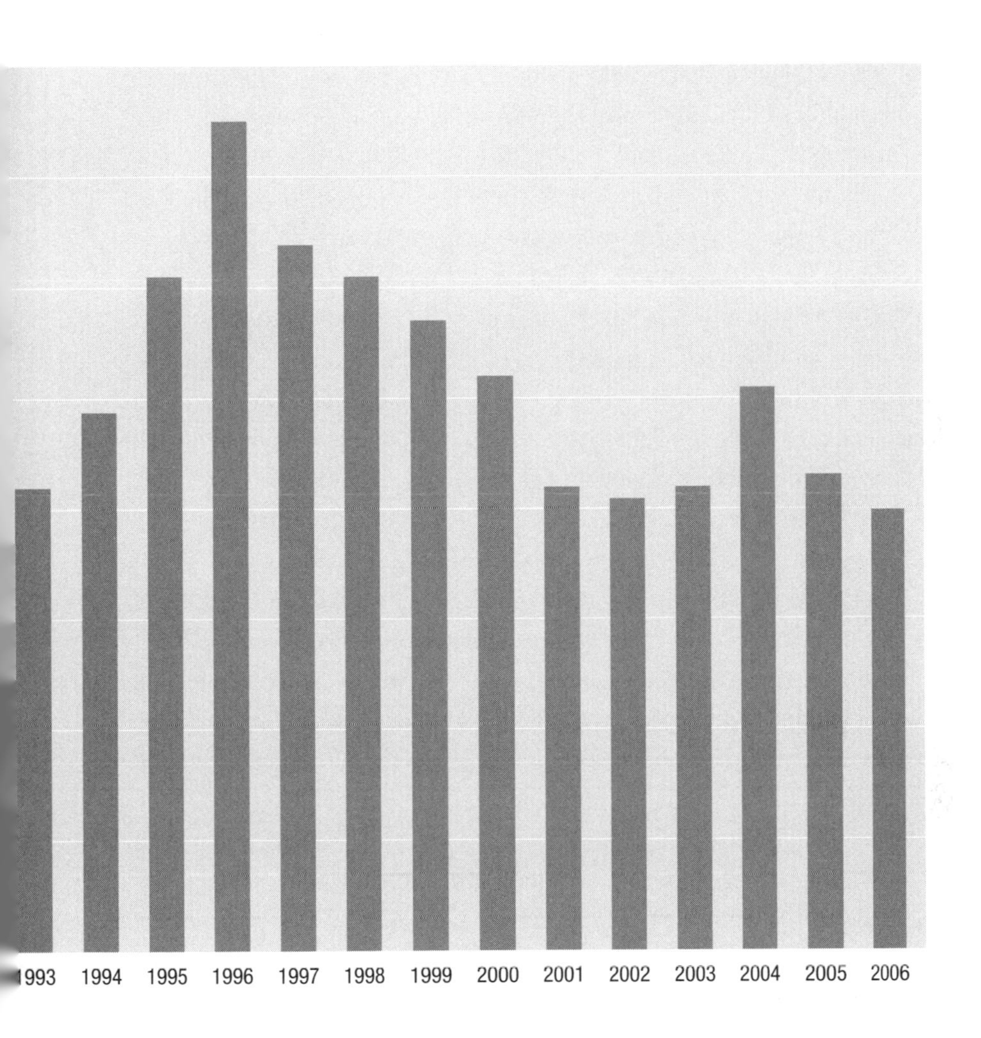

Denver-Julesburg operator, founded by Nick Sutton and Mike Higham. Subsequently, Petrie Parkman was an active comanager in many IPOs, including, among others, General Atlantic Oil (formed by Don Wolf), St. Mary Land & Exploration Company (Tom Congdon, CEO), Hugoton Energy (founded by Floyd Wilson), Flores & Rucks (founded by Jim Flores and Billy Rucks), and the spin-off of Union Pacific Resources from its railroad parent (under the leadership of Jack Messman, its CEO).

The Union Pacific Resources IPO was in many ways a highlight for Petrie Parkman's practice of executing in capital market financings. It built on a perspective dating back to my September 1971 analyst field (and railroad) trip to Omaha, Denver, and Corpus Christi. For both Paul Leibman and me, it levered off the investment research we had developed in the mid-1970s relating to the opening of the Overthrust Belt, first with the Pineview and Ryckman Creek field discoveries. These finds were followed relatively quickly by a series of other discoveries involving the Lodgepole, Clear Creek, Carter Creek, Painter Reservoir, Yellow Creek, Whitney Canyon, Anschutz Ranch, and Anschutz Ranch East fields. Union Pacific's land grant holdings in Wyoming and Utah encompassed significant portions of these Overthrust Belt fields with which we were familiar (see page 13). Furthermore, there was the achievement of having Petrie Parkman selected as a lead underwriter of this $900 million public placement of stock alongside Smith Barney, Goldman Sachs, and Credit Suisse First Boston. With this transaction, Petrie Parkman's capital markets strategy of being the most active comanager of choice in the E&P sector was now well recognized. Before the IPO cycle had fully run its course, Petrie Parkman was a manager of over twenty oil and gas IPOs, more than any other banking firm in this period. As will be related subsequently, this re-expansion of the E&P universe ultimately set the stage for yet another period of corporate consolidation as the petroleum economic cycle unfolded.

This series of successful E&P IPOs then gave rise to a broader acceptance of a wider range of energy-focused equity and debt underwritings. To participate in these opportunities, Petrie Parkman's analyst, Stu Wagner, developed research coverage of the midstream oil and gas sector, and our colleagues Lon McCain and Mike Bock developed a specialized corporate finance advisory practice focused on public and private placements of equity and selective debt transactions.

In addition to helping launch through the IPO process many of the new independent oil companies, Petrie Parkman embarked on a plan to become an active strategic advisor for companies looking to realign their positioning via merger, acquisition, or divestiture of selected corporate holdings. Beginning this initiative, in 1993–94, Jon Hughes and I undertook an advisory assignment for American Oil & Gas Corporation that ultimately involved the merger of the company into KN Energy, a Denver-based natural gas utility looking to expand by diversifying into the upstream exploration and production sector for oil and gas. Jon assembled a cadre of Petrie Parkman professionals to focus on developing the skill sets required for us to advise on this assignment and subsequently on various other corporate combinations. Following the American Oil & Gas transaction, our team worked with KN Energy's CEO, Larry Hall, on several M&A transactions, including its acquisition of the Midcon Pipeline operations from Occidental Petroleum (OXY). This transaction coincided with OXY's acquisition of the Elk Hills field in California from the U.S. government. Because the values of the two deals were within less than 5 percent of each other, OXY's pipeline sale in effect adroitly financed the significant bolstering of its U.S. oil and gas production footprint. This represented the first step in Steve Chazen's vision for a major repositioning of Occidental's U.S. upstream portfolio of assets. He followed it up with a similarly large acquisition from Shell Oil and BP Amoco of legacy oil properties in the Permian basin. Subsequently, we advised KN Energy on its merger with Kinder Morgan. This transaction provided much-expanded pipeline holdings for Kinder Morgan that could be utilized to tax-efficiently generate distributable income within the company's master limited partnership (MLP) affiliate. Thus, it catapulted Kinder Morgan into its leadership role in creating an energy income distribution category of securities generally analogous to the real estate income trusts.

Petrie Parkman was also a strategic advisor on a corporate merger that created a new operating entity, Cimarex Energy. This was achieved through the combination of Key Production with the upstream operations of the oilfield drilling company Helmerich & Payne (H&P). This transaction involved a tax-efficient structure known as a reverse Morris Trust. Key Production had been spun out of Apache Corporation in the mid-1990s. It was a vehicle that the late Mick Merelli, a former Apache president and chief operating officer, had fashioned into a streamlined and rationalized Midcontinent producer. Hans Helmerich, the CEO of H&P, approached Petrie Parkman to discuss his vision for divesting his upstream assets in a manner that would enhance value for his shareholders.

This strategy was all part of a bigger plan by Helmerich to focus the company back on its core business, whereby he envisioned the opportunity to implement a major upgrade to its fleet of drilling rigs. Realizing that the cyclicality of oil prices and consequently that of petroleum industry capital spending was a given, H&P had decided that a major modernization of the rig fleet was overdue and could provide a decided competitive advantage. Management opted to be the leader in accomplishing this utilizing new "Flex Rig" designs that would improve operating performance while dramatically reducing the time required to relocate and set up rigs on new locations. This innovation would yield economic benefits for the company's oil industry customers with lower overall drilling costs and improved drilling efficiency and accuracy in terms of staying "in zone," while reducing accidents. The market receptivity to the combination of Key Production and H&P's upstream position was positive from the onset of Cimarex's formation. Over the ensuing years, H&P shareholders have benefited from a manyfold increase in the value of their E&P holdings. In addition, H&P was freed to focus on the makeover and build-out of its new drilling rig fleet. This transformation is an outstanding example of the ability of a focused and properly incentivized private capital enterprise to regenerate its competitiveness in the dynamic and challenging energy markets. With this and other successful transactions by the early 2000s, Petrie Parkman had established a market franchise as a capital markets participant with focused petroleum investment research and multiple investment banking capabilities in upstream and midstream asset sales as well as corporate mergers.

The Second Oil Price Collapse, 1997–1998

> Old geopolitical grudges should not be allowed to fester;
> they need to be put to rest.

Having witnessed the extremely stressful events of 1986 and watched OPEC and the petroleum industry learn a series of painful lessons stemming from an oil price decline of more than 75 percent, I, like many others, thought a repeat price collapse of similar magnitude to be almost impossible or at least highly unlikely. What follows is a tale that indicates once again that widely presumed unlikely events can be more probable than many realize or for which they typically plan. Therefore, when such black swan events do occur, they are by definition often more impactful than most market participants expect.

In July 1997, the stage was set for the second oil price debacle when Asia's emerging market currencies began to experience major devaluations. These problems started in Southeast Asia with a currency crisis in Thailand triggered by the loss of confidence by holders of the country's large amount of foreign debt. They were further complicated by overextensions in real estate. The resulting malaise spread quickly to Malaysia and Indonesia. Before long, even the much more developed, and presumed stable, economies of South Korea and Japan became embroiled in currency market turmoil. In turn, these developments had profoundly adverse economic consequences that shortly thereafter led to a serious miscalculation by OPEC's collective leadership.

On November 29, 1997, the eleven-member OPEC group met at their offices in Vienna. In a remarkable example of the tendency of otherwise well informed and sophisticated human beings to deny reality, they voted to raise the oil production ceiling by over 2 million barrels per day at a time of shrinking global demand! This action appears to have taken no account of the unfolding economic downturn stemming from the spreading Asian currency crisis. I distinctly recall that my reaction to OPEC's announcement was, "What are they thinking?" and then, alternatively, "Do they know something I don't?"

By March 1998, the answer to that second question was a resounding no. The OPEC increase was, in fact, a serious misjudgment, and it had been made at exactly the wrong time. Consequently, as individual OPEC states attempted with limited success to push higher volumes of oil into stagnant or shrinking markets, price pressures were mounting. At a follow-up meeting in February 1998, OPEC ministers candidly acknowledged their error and announced plans to revoke the November 1997 increases in production quotas in a belated effort to stabilize oil prices.

Market observers were left to wonder whether these actions were too little and too late. In any case, clearly the new volume restrictions would need some time to take effect and undo the damage from the excess oil exports of the previous months. A few months later it was evident that the decisions announced in February were not having the desired effect. While good data tracking of oil exports by individual OPEC nations was then (and often still is) elusive at best, the somewhat successful efforts by Saudi Arabia, Kuwait, and some non–Middle East OPEC members to rein in production appeared to be largely offset by an ongoing rise in output from Iran. Thus, as the frontispiece indicates, each oil price rally quickly failed and led to new lows throughout 1998. I recall reports at the time indicating that this led to back-channel discussions between Saudi and Iranian officials and appears to have generated renewed, but still only soft, Iranian assurances of future efforts to comply with the February OPEC decisions. However, by August 1998, it was once again emerging that nothing substantial had actually changed. Another follow-up discussion between Saudi and Iranian representatives at last revealed what really lay behind the Iranian gamesmanship.

The Iranians finally conceded that they were not inclined to adhere to the new OPEC quota guidelines until there was a satisfactory resolution of their long-standing grievance over the relative size of OPEC export market shares following the Iraqi invasion of Kuwait in 1990. That event, which had caused the removal of well over 4 million barrels per day of Kuwaiti and Iraqi output from the markets, had necessitated Saudi Arabia's making up most of the loss by increasing its output from 5 million to 8.5 million barrels per day. At that time, the Saudis had assured other OPEC members that these actions were temporary and that following resolution of the problem, Saudi Arabian output would return to its previous lower levels.

Unfortunately, after the Gulf War the imposition of sanctions on Iraq had delayed the return of its production to market and thus enabled the Saudis to assert the necessity of keeping their output at the higher levels. Over the course of several years in the mid-1990s, this lack of promised realignment in market shares became a major source of contention between Iran and Saudi Arabia. With the ill-considered November 1997 move to increase OPEC output and the revelation by early 1998 of its harmful effects, the Iranians finally saw their chance to bring the issue of an adjustment in market shares to a head. By pretending to support the February decision but actually repeatedly doing just the opposite, the Iranians had adroitly positioned themselves to focus Saudi attention on the issue they actually wanted resolved. In sum, they demanded a more than proportionate reduction in Saudi output to correct what the Iranians viewed as an unjustified "market share grab" coming out of the Gulf War crisis.

With the issue finally framed, a series of negotiations over another five or six months as well as the motivating effect of oil prices plummeting to levels near single-digit dollars per barrel were needed for an acceptable outcome to be formulated. While there were distinct differences in the drivers causing this latest oil price collapse, the entire experience amounted to an eerily similar replay of the oil debacle of 1986. This time it was only when Saudi Arabia finally agreed to consider significant reductions in its output that negotiations got under way with the Iranians. Even then, those efforts were further complicated and thus postponed by a crisis in August 1998 involving the Russian default on its international debt obligations. The knock-on negative effects of this event proved to be exceptionally widespread as well as destabilizing to the confidence of financial and oil markets. Long-Term Capital Management (LTCM), a large and highly leveraged hedge fund based in Greenwich, Connecticut, found that the utility of its historical correlations to determine numerous macro financial bets were breaking down. The resulting financial losses were disastrous because of the fund's extremely high level of leverage. The situation was further complicated by the uncertainties associated with the Russian debt default. This in turn roiled the major debt and equity markets globally, as it became evident that a painful bailout of LTCM by that firm's Wall Street banking counterparties would be required.

At long last in March 1999, Saudi Arabia agreed to accept a quota reduction to 7.4 million barrels per day, and OPEC was then able to formulate production adjustments by other OPEC members as well as selected non-OPEC producers. The total reductions of 2.1 million barrels per day provided for a credible plan to restore stability to oil markets. While markets and observers continued to question whether these moves would work, it was now arguable that there was discernible light at the end of the long tunnel. In a call remarkable for catching the oil price bottom the wrong way, the usually much more astute *Economist* magazine published a cover story entitled "Drowning in Oil" in an early March issue that predicted that oil prices could fall further to a level below $6 per barrel. On seeing this article, I remarked to colleagues that it had the potential to rival *Business Week*'s August 1979 now infamous "wrong call" cover story that featured "The Death of Equities" not long before the onset of history's greatest bull market run. Both of these articles failed to recognize that it is often darkest before the dawn. However, much to its credit in terms of exhibiting intellectual honesty, the *Economist* published a second article in December entitled "Goofs: We Woz Wrong," acknowledging its error and analyzing why it was so mistaken. By that time, oil prices rallied from the $12 per barrel level to nearly $20 per barrel.

As the ultimate reconciliation efforts were playing out, I was in Saudi Arabia in February 1999 in connection with Petrie Parkman & Co. becoming engaged as an advisor to the Kingdom's foreign minister on that country's natural gas initiative. Thus, I had the opportunity to witness at closer hand the Saudis' efforts to achieve closure on this issue. As subsequently recounted in local press reports, the final agreement with Iran came out of a call arranged by the Saudi foreign minister Prince Saud al Faisal, between then–Crown Prince Abdullah and Iranian president Khatami. For me, the lesson learned was that old geopolitical grudges between countries and differing cultures should not be allowed to fester. They need to be identified and dealt with on a timely basis to avoid incurring serious collateral damage.

Saudi Gas Initiative

Culture matters, especially in the international sector.

In late October 1998, there was a widely reported tea party hosted in Washington, D.C., at the residence of the Saudi ambassador to the United States, Prince Bandar. Attending were the Saudi foreign minister, Prince Saud al Faisal, and Crown Prince Abdullah, who had been effectively running the Kingdom following King Fahd's debilitating stroke in 1995. The invitees consisted of senior representatives of seven of the largest international oil companies (Exxon, Chevron, BP Amoco, Royal Dutch Shell, Texaco, Mobil, and Total). The crown prince opened the meeting with a surprise announcement that the Saudis had decided that they were ready to consider opening the country to "selective" new investment in portions of its energy-related infrastructure.

While there was a noteworthy degree of vagueness to this assertion, the very idea caught many of the attendees off guard. What they initially thought would be primarily a social event had taken on the potential to be historically momentous. Not since the nationalization of Aramco (with negotiated compensation) in 1978 had there been even a hint of welcoming foreign capital back into the heart of the Saudi energy complex. The notable, but still relatively minor, exceptions included several downstream refining and petrochemical joint ventures and Texaco's participation (through its acquisition of Getty Oil) in the Saudi/Kuwaiti Neutral Zone redevelopment after the extensive damage of wells by Iraqi soldiers during the Gulf War of 1991. Because the latter involved developing heavy oils akin to those of Texaco's California operations, the company clearly brought particularly relevant experience with the requisite techniques to deal with this difficult heavy gravity crude oil.

In retrospect, the scope of the Saudi concept apparently was not very thoroughly discussed in this initial meeting. However, there was the implication that investments in the natural gas sector would be welcomed, including upstream development (that is, drilling of natural gas wells). In any case, the invitees departed the meeting, and each company proceeded to form a task force to be ready to negotiate for projects when the Saudis would actually issue invitations.

This all occurred while Jim Parkman, Randy King, and Jon Hughes were wrapping up Petrie Parkman and First Boston's engagement to sell the Elk Hills Oil field in conjunction with the Clinton administration's decision to privatize the Naval Petroleum Reserve in California. As press accounts appeared about the tea party surprise announcement regarding a possible Saudi reopening for foreign investments, Jim Parkman and I observed that some and maybe much of what we had learned regarding the Elk Hills governmental divestiture process could be applicable to an engagement with the Saudis. However intriguing the thought, though, we were not inclined to pursue it at that time because of numerous opportunities and commitments we had closer to home.

That view changed rather suddenly in February 1999 when I received a call from an oil company CEO who gave me a heads-up that he had recommended Petrie Parkman & Co. for a role as a co-advisor (alongside Morgan Stanley) to a senior official of the Saudi government on that country's plans to expand development of its conventional, prospective natural gas resources. Four days later, I made a quickly arranged trip to Riyadh (compliments of my visa application being expedited by the Saudi consulate). There I learned that the project would entail basic technical and economic assessments of a range of options that focused on the potential for natural gas projects to generate electricity, desalinate water, and integrate new methane supplies into petrochemical processing facilities. The engagement called for providing advice on strategies for upstream projects to ensure adequate availability of domestic gas supplies for such facilities. The initial plan envisioned a series of new joint ventures with international oil companies to explore for and develop natural gas projects as well as plans to integrate new supplies into downstream projects to meet Saudi Arabia's basic domestic needs for water, electricity, and petrochemicals.

Jim Parkman and I made another trip to Jeddah, Saudi Arabia, later that spring to better understand the scope of the assignment. During this visit, we encountered a situation that underscored the importance of having an adequate degree of leverage whenever one is negotiating a deal. After several days of work, we noticed that there was a Bedouin Arab offering a ride (and a Polaroid photo) on a camel in a park across the street from our conference center. Jim and I decided that we would like to take him up on the offering. In the negotiation over the price of his service, we were able to achieve a significant reduction in the proposed cost of our rides. However, when I took the first ride around the park, he stopped the camel halfway and explained that that was all that I had paid for and that I would have to walk back and that there would be no photo. When I capitulated, we returned, and he took the same tack with Jim. In the end, we paid what he had originally stipulated as his price. Ever since this incident I have been sensitive to the moment one loses negotiating leverage in doing any deal.

Following our visit to Jeddah, we hired Mark Sooby to ensure a continuity of focus on the project. We had known Mark previously in connection with another engagement with a private independent. Most recently, he had headed up ARCO's team preparing for the Saudi gas initiative until the acquisition of ARCO by BP Amoco. Because BP Amoco had its own team in place for the Saudi project, Mark joined Petrie Parkman. He came to us with a high degree of familiarity with the Saudi goals, key personnel, and issues. Jim Parkman, Mark Sooby, and several other colleagues then worked on this effort over a four-year period, after which it was taken over by Saudi

Aramco. At that point, Petrie Parkman and Morgan Stanley's advisory role was terminated.

The project subsequently appeared to lose momentum and did not achieve the scope of impact that had been first envisioned. This was mostly because of a lack of exploratory successes in the regions awarded to the joint venture participants. These drilling disappointments were probably attributable to several factors. First, the areas in which new concessions were awarded largely involved new basins with true wildcat risks. Thus, even in Saudi Arabia the probabilities of exploratory success were inherently low. Second, most of the companies with whom the deals were ultimately struck were new to Saudi Arabia. For many reasons (including, among others, low expected rates of return), the traditional U.S. super majors with in-depth historical knowledge of Saudi Arabian geology and technical operations did not feature among the finalists. Instead, the common approach of the newly entering players, including Chinese, Russians, Australians, and others, appeared to be marked by a willingness to gain a foot in the door with low expectations for financial returns. These partners likely had hopes of positioning themselves for better deals at a later point. When the lack of prospectivity in the initial concessions became evident, the odds of follow-on opportunities diminished, and the sense of commitment undoubtedly faded given the prospect of high capital outlays with real "dry hole" risks. Nonetheless, the process did highlight a Saudi need that continues to the present time to develop new natural gas supplies. On that point, given recent technical progress globally involving development of unconventional shale gas resources, new internal Saudi initiatives by Saudi Aramco still seem possible and potentially promising.

Creating Super Majors

Merger-Mania II: A case study in how actions
and events beget additional actions and events.

Closely intertwined with the foregoing described events relating to the second oil price collapse, as well as the hoped-for Saudi reopening to foreign investment in energy projects, was the onset of a second wave of corporate merger consolidation at the very top of the Western Hemisphere petroleum sector corporate hierarchy. Over slightly more than a three-year period, it would involve either as acquirors or as acquirees all but one of the companies represented at Prince Bandar's tea party to launch the Saudi gas initiative. This consolidation became the largest ever of its kind. In terms of total market capitalizations involved, it surpassed by a wide margin both the size and the implications of the previously described M&A wave of 1980–85. It also triggered second-order (knock-on) effects that precipitated other large subsequent transactions. When it was completed, Exxon and Mobil would be fused into a single enterprise, as would BP Amoco and ARCO, as well as Chevron and Texaco (see figure on pages 52–53). In turn, Total would be moved, possibly for defensive reasons, to acquire Petrofina and Elf. In addition, after having rather passively accepted these major transactions for U.S.-based companies, the Federal Trade Commission (FTC) would finally draw its line in the sand over BP Amoco's acquisition of ARCO. Before permitting that merger to occur, the FTC demanded the complete divestiture of ARCO's North Slope assets to ensure that BP would not become too dominant a West Coast domestic producer. This confrontation then led to competitive bids for ARCO's Prudhoe Bay assets by both Phillips Petroleum and Conoco. Given its successful bid, this transaction bolstered Phillip's negotiating leverage in its ultimate merger with Conoco a year and a half later. Thus, in a matter of only about forty months, twelve large-capitalization petroleum enterprises had been reduced to only five freestanding but even more gigantic enterprises. Only the Royal Dutch Shell group was without a partner as the music stopped.

Daniel Yergin has articulately reconstructed the behind-the-scenes particulars of the interpersonal communications that drove this unprecedented wave of mergers in his book *The Quest*, a comprehensive update to his earlier Pulitzer-winning account, *The Prize*. What emerges from his narrative is a sense of how tightly linked are the perceptions of how each CEO player's actions often directly affected the mindset of various competing managements. Thus, each transaction announcement motivated the decision framework of one or more of the other key industry participants. As Yergin details, when Mobil insisted on a deal-killing premium to merge with BP, John Brown, BP's CEO, instead decided to approach Amoco's CEO, Larry Fuller, who was open to doing a deal. It was announced in early August 1998. This new valuation and the resulting insight to the relatively benign FTC regulatory stance regarding big oil mergers in turn revitalized Exxon's until-then-stalled discussions with Mobil. Just before year-end 1998, Exxon and Mobil announced their agreement to merge. Only a month later, ARCO's CEO, Mike Bowlin, chose to initiate discussions with BP Amoco that ultimately turned into a 1999 April Fools' Day public announcement of BP's acquisition terms. Notwithstanding the date of the press release, this ARCO decision was puzzling, but it was not a spoof.

Precisely how much the U.S. consolidation wave influenced the strategic thinking of European companies is unclear. However, it is interesting to note that Total proceeded to roll up Petrofina in late 1998. It then turned its attention to Elf Aquitaine by mid-1999, consummating that merger with governmental approvals in September that same year.

Back in the United States, recognizing both the handwriting on the wall and the logic of a combination given their shared ownership of Caltex, Texaco's CEO, Peter Bijur, reached out to Chevron that same spring. This one took longer to consummate, with serious negotiations starting in the spring of 2000 and the public announcement coming that fall. Finally, Conoco and Phillips announced their agreement to become hitched in late 2001. At the end of this consolidation cycle, only Royal Dutch Shell stood apart from this incredibly compressed period of merger transactions. Toward the end of this era, Shell did launch an ultimately unsuccessful unsolicited offer for Barrett Resources, a leading Denver-based independent producer of predominantly Rocky Mountain–sourced natural gas.

On reflection, several factors appear to have contributed to the mindsets of the CEOs and boards of these companies during this period. Clearly, the second oil price collapse provided a depressing and challenging backdrop for conventional upstream development. This undoubtedly motivated managements to consider how cost efficiencies could be achieved by spreading overhead over a larger base of operations. Also (as described in the next chapter), there was renewed concern about the looming prospect of a global peak in oil output. Thus, the old adage about finding cheaper oil on Wall Street was again receiving greater strategic consideration. Finally, I suspect that the diminishment of the initial high expectations for a Saudi reopening via its natural gas initiative may have played a small but noteworthy role in influencing some of the later transactions in this cycle.

In any case, the landscape was significantly changed. For the moment, the number of large-capitalization public vehicles for institutional investors was cut to less than half that of the period before 1998. This consolidation began a trend of freeing up technical and managerial resources for new entrepreneurial endeavors. As the full integration plans associated with these mergers were executed, talented and experienced personnel began to migrate to operating platforms where new private equity funding began to be applied to new ideas for upstream petroleum development. This was capitalism at work, a pattern that would gather additional momentum during the first half of the next decade.

9/11, 2001

The previously unthinkable becomes the new reality.

In early 2000 in connection with Petrie Parkman's Saudi Arabian gas advisory assignment, Jim Parkman came across an indication on the Internet of Osama bin Laden's now-infamous fatwa calling for Muslims to be ready to kill Americans wherever they might come across the opportunity to do so. Accordingly, we directed our employees working on the project to lower their profile when working in Saudi Arabia. Thereafter, they stayed in Western hotels with better security, and we discouraged employees from spending time out in the economy or countryside. The appropriateness of these and other measures became clear when investigations revealed the role that Osama bin Laden played in the Kenyan Embassy bombing. These findings were detailed in a book entitled *Bin Laden: The Man Who Declared War on America,* by Yossef Bodansky.

Accordingly, when the first plane hit the North Tower at the World Trade Center on September 11, 2001, I was immediately skeptical of news coverage indicating that it may have been an accident. Instead, I instantly suspected it was another strike by the Al Qaeda terrorist organization, which had already been implicated in the 1993 World Trade Center bombing incident. It was not long before the second plane hitting the South Tower confirmed my suspicions. That morning Jim Parkman was returning to the United States from Saudi Arabia. He had taken a Lufthansa flight from Riyadh to Frankfurt. Having connected to a flight for Houston, he was over the Atlantic when the planes hit the World Trade Center. His plane was diverted to Gander, Newfoundland, along with dozens of others. The hospitality of that city, which had about tripled in population overnight, was extraordinary. Jim spent the next several nights sleeping on the floor of a church. A day or so into this crisis, Jim called and asked that we arrange for a private plane to come pick him up when the ban on incoming flights to the United States was lifted. When we asked why the German plane would not be available, he explained that because of Arab passengers, the aircraft would be returning to Frankfurt. When the ban on flights was lifted, we sent the plane, but they still had to fly over Canada west of Chicago before entering U.S. air space to go to Houston.

Because so much has already been written about the events of 9/11, I will not recount the wrenching details here, with an exception noted in the paragraph below. Instead, now with the benefit of over a decade's separation, it is appropriate to focus on how this tragedy accelerated a trend that was already under way involving the reconfiguration of geopolitical relationships that have a direct bearing on twenty-first-century petroleum positioning issues for the United States and many of its Western allies.

Regarding the exception I mentioned above, on September 11, 2001, I had been planning to attend an oil analysts symposium conducted by Scott Rees and Tom Tella, petroleum engineers with the firm Netherland, Sewell & Associates, on oil and gas property valuation being held on the forty-fourth floor of the North Tower of the World Trade Center. However, what proved to be a most fortuitous schedule conflict because of a client meeting in another city had prompted a change in my travel plans a week earlier. What follows is a stirring firsthand account by my friend the late Al Anton, a partner with Carl H. Pforzheimer & Co., on the travails of those attending that meeting hosted by the National Association of Petroleum Industry Analysts and their close calls while successfully navigating their escape to safety:

A Day Like No Other

By Albert J. Anton, Jr.

September 15, 2001

The passage of time distorts perceptions and fogs the memory. I thought it worthwhile, therefore, to record for friends and relatives my impressions of that terrible day, September eleventh of 2001:

The day started out rather routinely, with the exception that I would take the 7:30 AM train to Hoboken rather than the 7:14 "Midtown Direct" to Penn Station, New York. Instead of going to my office on Madison Avenue at 59th Street, I would go downtown to the World Trade Center for a seminar on the esoteric subject of oil and gas reserve valuation.

At Hoboken, I transferred to the PATH train, which would bring me in fifteen minutes to the cavernous station deep below the twin towers of the WTC. I had made the journey thousands of times—when my office was downtown—but had only done it a few times in the last ten years. I reported to the security desk at One World Trade Center at about 8:30 AM. Security was rather rigorous and, having neglected to bring my pass as a member of the New York Society of Security Analysts, I had to wait until the guard produced a duplicate—dated 9/11/01—from the center's computer system. I proceeded to a bank of elevators, where I presented my pass to another guard and entered the oversized express elevator for the 44th floor. There, at the NYSSA meeting room I met some friends and settled down to a cup of hot coffee.

The seminar was to begin at 9:00 o'clock, so at about 8:45 I suggested to my friend, Richard Cohen, that we take our coffee to the north windows to enjoy the view. Suddenly there was a roar of engines; Dick said, "A plane, and he is flying low!" By the time I looked up I saw only a blur above, and then the building shook, making it hard for me to maintain my balance. Then, all was quiet. We saw no flames, no smoke, no falling airplane parts, only fluttering shards of glass and scraps of paper. My friend said he thought the plane was a corporate jet, not the massive 767 it turned out to be. Others agreed. (I realized later that, at full throttle of perhaps 500 miles per hour, the plane was likely a half mile away when you could last focus on it.)

Everyone remained calm—too calm. We spent some minutes talking over what had happened and wondering what to do next. No alarm sounded. The emergency strobe lights on the ceiling remained dark. The elaborate speaker system that had been installed after the 1993 terrorist bombing was silent. I phoned Sara at 8:55 and left a message that a plane had hit us but that we would be all right. (Just in case these would be my last words to her I told her that I loved her.) I then called my office and had the Dow Jones report of the crash read to me. There was no mention of a terrorist motive. Only when we smelled smoke in the hallway did we decide to take the stairway down. This was about twelve to fifteen minutes after impact.

We took the "C" stairway, a surprisingly constricted egress for so large a building. The going was slow, owing to the large number of people, but no one panicked. Despite eye-burning smoke, a spirit of comradeship developed. Some wondered aloud when they would be able to return to their office to pick up briefcases and other items left behind. About halfway down we were asked to stand to the left to allow a group of 20 firemen to go up to the fire. They were carrying hoses, axes, chemical tanks and other heavy equipment. We all wished them well and I uttered a silent prayer. I did not know then that we were watching them go to their deaths.

At the 20th floor the smoke cleared and going was faster. Again, we were passed by security personnel on the way higher. Around the fifth floor we encountered water cascading down the stairs, making footing difficult. Finally, we emerged into the spacious upper lobby, always so bright and striking. I had long considered it the best architectural feature of the center. But now it was dark and foreboding. The great central plaza was a war zone, with rubble everywhere and bodies of the fallen. It was only then that I fully realized how precarious was our situation.

We were led to the right, around the building's core to an escalator, where we walked down to the lower lobby—which was flooded with three inches of water. Security personnel in anguished voices directed us out into the shopping concourse beneath the plaza. The sprinkler system showered us from above, while water on the terrazzo floor made going difficult. We were directed straight past a line of stores and banks, then to the left before the PATH escalators, then turning right toward the Church Street entrance, where we ascended an escalator to the street. Firefighters outside the center directed me across Church Street and up Fulton Street, with St. Paul's churchyard on my left. I turned to look at the twin towers, and, for the first time, realized that they both had been hit, the South Tower having been impacted while we were deep in the North Tower stairwell.

I crossed Broadway where a large crowd was watching the conflagration. Someone said, "There goes another one!" I looked around to see a man jumping from a high floor, soon followed by another. I couldn't watch and walked ahead, praying and nearly crying. (I'm sure that atheists were praying that day and I saw many a grown man weep.) Continuing on Fulton Street toward Nassau Street, I heard a roar and turned to see a massive cloud of smoke and debris rushing toward us. I soon learned that this was the collapse of the South Tower. I was out of the North Tower no more than five to seven minutes, and have no way of knowing if my exit would have been blocked by the south building's collapse. I picked up the pace and ducked into the subway station at Nassau Street, only to find that trains were not running. A block further, I tried the subway at William Street, where trains were operating but not stopping. I was further east near Gold Street, when another roar and cloud of smoke and dust signaled the fall of the North Tower—which I had left less than 45 minutes before.

I walked north, almost dazed, though the Southbridge Towers, under the Brooklyn Bridge and up St. James Place into Chinatown. Thousands of people were walking across the bridge to Brooklyn, both on the pedestrian walkway and on the vehicular roadway. I worked my way up Mott Street to Canal Street, where a number of people remarked on my dust-covered clothing. Merchants had placed television sets in their doorways, and cars and trucks were stopped with radios blaring the news of the tragedy. People were lined up six to eight deep at pay phones. At Canal Street I walked a block west to Mulberry Street, which was decked out for Little Italy's San Gennaro Festival (which, of course, would be cancelled). Then via Broadway to Ninth Street and over to Sixth Avenue, where I found that the uptown division of PATH was not operating. I then walked west to Horatio Street to the apartment of my friend, Sal Ilacqua, who was also to attend the seminar. Happily, Sal had survived and we were later joined by Barry McKennitt, Wayne Whipple, and my partner, Frank Reinhardt.

During all this time Sara did not know if I were dead or alive. My son,

Chris, had come over from his law office and they were both in a panic. I phoned home at 10:45 from a pay phone in Chinatown, but the circuits to New Jersey were overloaded. A computer voice asked if I would like to leave a message (for a fee) and I did. But that message was not delivered until later that afternoon. In Louisville, my daughter, Claire, went to the office of her husband, Barry Whaley. There, at about 12:15 PM, they received word from my office of my safety. They, in turn, called Sara and Chris with the news. My son, Tom, was teaching in Irvington. He was preparing to come to our house when he learned that I was safe.

The numbers of survival stories are endless. Frank Reinhardt had gone downstairs in search of a breakfast Danish just about the time the plane struck. When he was barely a safe distance from the elevator it burst in flames from jet fuel that had poured down the shaft. Sal Ilacqua was on his way to the building when he met a crowd running in the opposite direction. He walked north on Church Street, where he saw an engine pod from the second plane fly over his head and land in the street. Several people were hit by the debris and killed. Another friend, Merz Peters, was walking across the central plaza when the first plane hit. He survived by diving beneath a granite bench.

People ask me if I have slept in the four nights since the disaster. I have had no problem sleeping, but when awake, I have been fixated on two thoughts. First, the fact that I was quite literally at ground zero in the attack: on the north face of the North Tower, albeit 50 stories below the point of impact. If the plane had come in lower or if the stairway was blocked or doused with jet fuel, I would not be here to tell my story. Second, the image of young firefighters and security people heading up the stairs as we were going down has given a new meaning to duty and honor and responsibility. The picture of these men—in all likelihood going to their deaths—has been seared in my memory. I can only think of Jesus' words in John 15:13: "Greater love hath no man than this, that a man may lay down his life for his friends." The hate that inspired the raid has been eclipsed by the love and compassion and devotion to duty that has characterized New York and, indeed, all America in the past few days. If we can retain this spirit, all the loss and suffering will not have been in vain.

———

Whenever I am assessing various aspects of the evolving twenty-first-century geopolitics of energy, I often recall Al Anton's stirring account of this tragic, cataclysmic event.

Geography Matters

It is all about location, allies, and competitors.

When I was growing up, I had a blind great-uncle who taught geography at a private school near Boston. His blindness was due to an inherited disease called retinitis pigmentosa. It amounts to gradual deterioration of the retina and is detectable at a relatively early age, so by his late teenage years, my great-uncle knew that he would be blind by the age of forty. With this in mind, he educated himself to become a teacher of geography and practiced that profession into his seventies, having memorized maps of the world back when he still could see them. He was accomplished enough that he held his position at the Fessenden School drilling the importance of geography into students, including, among others, Howard Hughes and Ted Kennedy. He often tested me and my brothers and cousins in the same manner as his students at weekend family get-togethers. After becoming an oil analyst, I came to appreciate those sessions with my great-uncle because of the solid foundation they provided for my understanding of how the geopolitics of oil can be shaped by key geographical considerations. In this respect, I have concluded that in the aftermath of the 9/11 event, a shifting focus of attention eastward into Asia and other parts of the Eastern Hemisphere by Saudi Arabia and other Arab exporters carries with it the potential (and even likelihood) to diminish U.S. influence in the region. A *Wall Street Journal* editorial appearing just after the September 11, 2012, assassination of Christopher Stevens, the U.S. ambassador to Libya, aptly called it "the new world disorder." As described below, that phenomenon is likely what we are now witnessing.

To fully appreciate the importance of some of the post-9/11 changes in the global geopolitical picture, one should reflect on the remarkably strong ties that were enjoyed by the United States and Gulf Cooperation Council countries (especially, Saudi Arabia and Kuwait) from the end of World War II up to the beginning of this new century. It is difficult to overstate the depth and scope of the relationships involved. In particular, the "front yard" of Saudi Arabia in effect looked west across the Mediterranean into Europe and beyond to the United States for its economic prospects as well as partnerships that would help ensure its security interests. The growth in demand for Saudi and Kuwaiti oil exports was a key factor reinforcing these relationships. Equally important, the rising wealth of the oil-exporting nations gave rise to growing demand for manufactured goods and services from the West. It was an almost ideal win-win international trade pattern that extended over five decades. To be sure, there were periods of strain in the relationships, sometimes caused by conflicts involving Israel's ongoing occupation of Arab lands stemming from the 1967 Six-Day War and other episodes of violence in the region. However, these events were relatively brief interruptions in the long-term expansion of bilateral trade.

With the perspective of the decade now passed, it is becoming clear that the favorable and predictable half-century trend in international trade is fairly rapidly giving way to new patterns of economic and diplomatic activity. These changes would probably be happening in any case, as growing Asian petroleum consumption is becoming a much larger driver of demand in the global picture and as Chinese manufacturing capabilities approach world-class status. The 9/11 tragedy marks a clear inflection point in this pattern of activity. It set decisively in motion forces that accelerated the pace of the transition under way.

Post-9/11 there has evolved a geopolitical power triangle involving Russia, China, and Iran. This is a de facto set of relatively new economic relationships that have emerged. There is no unified trilateral agreement codifying this power triangle. Nonetheless, it is arguably one of the most important new geopolitical realities to emerge since the end of World War II. The root of this new reality rests on several fundamental considerations. First and foremost, almost a quarter of China's growing oil demand is provided by Russia and Iran. This is a level that could well expand further over the balance of this decade. Second, the Moscow-to-Tehran axis of the power triangle reflects shared energy goals, such as Moscow's desire to export its nuclear power technology and Iran's need for the same. Evidence of the reality of the power triangle that I am describing is not always obvious, but it is nonetheless a rather persuasive construct for understanding today's geopolitical realities. Most notably, there are the on-again/off-again positions of China and Russia in the U.N. Security Council deliberations regarding proposed sanctions of Iran. As Iran's nuclear processing capabilities have progressed, possibly (or probably) as part of a potential nuclear weapons program, Russia's and China's tendencies to alternate between tepid support of weak sanctions and threatened vetoes of strong ones speak volumes about both the reality and the effectiveness of this power triangle.

The influence of this now-evident power triangle on Saudi Arabia is particularly noteworthy. One way to think about this is that the "front yard" of Saudi has been shifted from historically facing West to now facing East across the Gulf. This is not because the Saudis want to make that shift but rather because they have to do so defensively to look after their strategic national interests in the region. The United States and other Western consuming nations are no longer the key, or even noteworthy, sources of growing oil demand. In fact, the United States is now on a path to become a longer-term shrinking source of demand for Middle East and even Saudi oil. In contrast, China, India, and other Asian countries are experiencing growing demand for this oil.

Because the Saudis have to deal with the issues posed by the Beijing-to-Tehran axis, they are developing a more direct economic and commercial engagement with China. In addition, they are already beginning to forge a closer alliance with Turkey, perhaps in an effort to form an axis of leadership in the region to address local challenges such as those arising from the escalating civil war in Syria. This does not necessarily mean an end to Saudi/American relations, especially given a longstanding tradition of effective military cooperation and trade. However, it does suggest that the old "special" relationship is diminishing somewhat as it inevitably becomes more multilateral. Accordingly, it presents a much more complicated situation with attendant pressures on the U.S. president and other political leaders in Washington to use diplomatic creativity in adapting to and even capitalizing on this reality.

Regarding the complexity of this situation, there are the beginnings of a second power triangle in the region as well. This one involves an axis from China to Iran and Iran to India. Completion of this triangle's third leg (India to China) is more decidedly problematic, involving as it does the

EVOLVING GEOPOLITICAL CONSIDERATIONS

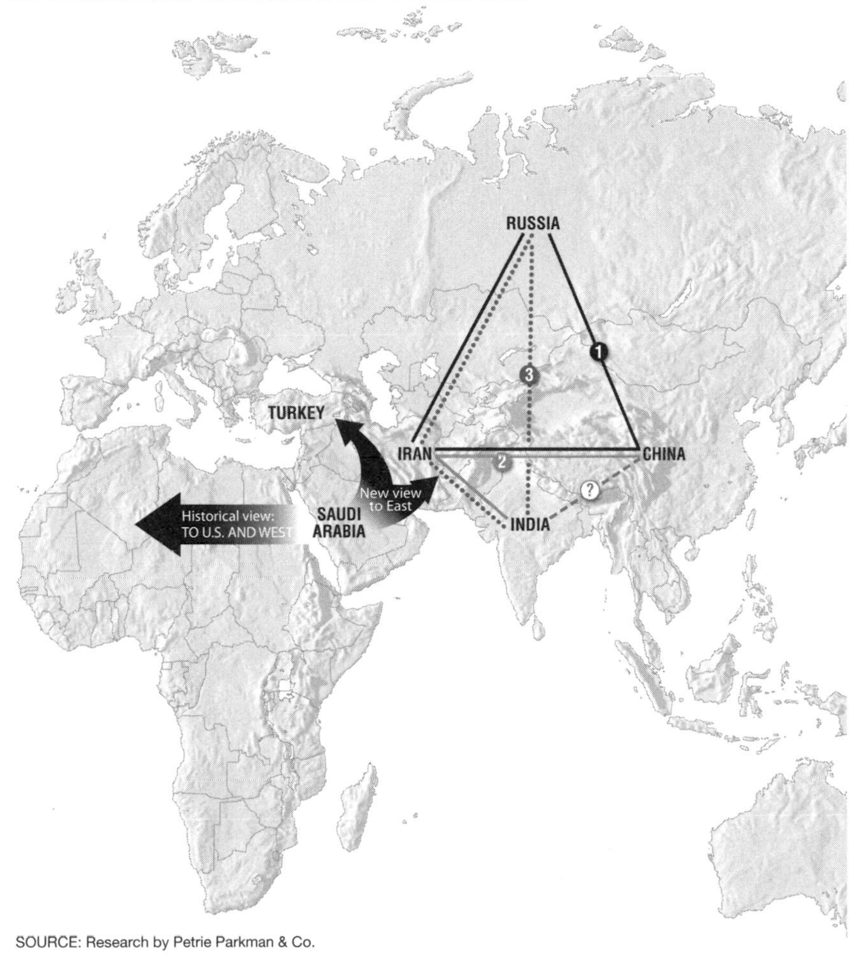

SOURCE: Research by Petrie Parkman & Co.

political obstacles to putting in place an axis across the Himalayas from India to China. Given a long, troubled history between these two countries, whether and how such a third leg will be established remains to be seen, but in the aftermath of the 2010 visit of the Chinese president, Wen Jiabao, to India, it should not be lightly or automatically discounted. For example, in June 2012, China and India announced the signing of a new joint energy agreement, signaling plans to coordinate and cooperate in upstream joint ventures to secure new oil and gas supplies. This cooperation stands in stark contrast to their previous adversarial roles as competitors in such endeavors over the past decade or more.

There is also a somewhat more remote, but not entirely dismissible, possibility of a third (in fact, overlapping) power triangle involving Russia, Iran, and India. The basis for this alignment stems from the common interests of these three countries in the long-proposed "pipeline for peace." Russia has an interest in promoting the export of Iranian natural gas by pipeline eastward to Pakistan and ultimately to growing, potentially large markets in India. This project would afford Russia the dual benefits of a chance to involve its large state-owned gas company, Gazprom, in a developing major new pipeline infrastructure projects while further helping to protect and perpetuate Russia's already strong position in supplying European natural gas markets. In March 2013, Iran and Pakistan announced an agreement to move forward on the Iran-to-Pakistan leg of this system. *Economist* magazine has noted many outstanding obstacles to making this project a reality. Nevertheless, it deserves ongoing monitoring as an indicator of shifting commercial alliances in the region.

The implications of these three interrelated power triangles for the future U.S. role in Asia are especially noteworthy. Thus, against the backdrop of the now-completed U.S. pullout from Iraq as well as the accelerated plans for withdrawal from Afghanistan, a reconfiguration of the traditional spheres of influence through the Middle East and South Asia is becoming evident. These developments are all reminiscent of the United Kingdom's decision in the late 1960s to early 1970s to abandon on a phased basis its longstanding security enforcement role for its postcolonial relationships "east of Suez." That process necessitated a significant step-up by the United States to fill the void. This time as the United States phases down, precisely how the vacuum will be filled is less obvious. In any case, it is likely to involve players from one or more corners of each of these power triangles. All in all, a series of rapidly evolving Middle East/Asian relationships are likely to provide a classic study in "why geography matters."

Chapter 6

GAME CHANGERS II

Merger-Mania III

Periodic consolidation is integral to the evolving natural economic order of private-sector petroleum enterprises.

By 2001, the recovery in the commodity and securities markets and the return to a business expansion mode by independent E&P companies was well advanced. Thus, many CEOs and boards of directors began to examine more seriously the option of sale or merger versus pursuing going concern strategies. The stocks of the independent S&P E&P universe had appreciated by 15 percent and 82 percent in 2000 and 2001, respectively, fully recovering the losses of 12 percent and 40 percent in 1998 and 1999. After some backing and filling in 2002 and 2003, this set up what would become the sector's best performance in well over a decade with appreciation of 27 percent, 34 percent, and 68 percent in 2004, 2005, and 2006, respectively. In addition, there was the onset of a war against the Taliban in Afghanistan post-9/11 and the looming prospect of further confrontations with Saddam Hussein in Iraq. Both of these developments contributed to the more than doubling in oil prices from the late 1998 low of about $10 per barrel and added further motivation for sellers to consider the sale or merger option.

The desire of larger producers and integrated oil companies to begin reestablishing or enlarging strategic positions in North America was becoming a factor. More than a decade earlier, some of these companies had all but abandoned onshore upstream petroleum exploration and development efforts in the United States. In some cases, this withdrawal even occurred in Canada following the oil price collapse of the mid-1980s. In the interim, independents had begun to pursue new approaches to unlocking economic hydrocarbon resources across a wide range of North American basins.

Typical projects included new natural gas development involving coal bed methane and tight gas formations in the Rocky Mountains, East Texas, and Midcontinent basins. Independents were also beginning to experiment with innovative techniques to find and complete development of resources in unconventional reservoirs and were having noteworthy success in building substantial levels of new economic value. The first sign that companies "up the food chain" from the midcapitalization E&P companies might have renewed interest in such exposures came in the first quarter of 2001 in the form of an unsolicited tender offer at $55 per share by Shell Oil for the shares of Barrett Resources Corporation. The latter had been founded in the late 1970s by Bill Barrett, a highly respected Rocky Mountain explorationist who over his previous career had several giant oil and gas discoveries to his credit.

I first met Bill in the mid-1970s when he was with Rainbow Resources, a small Denver-based independent that had shrewdly assembled and promoted to major oil companies large exploratory land positions throughout Wyoming, Montana, North Dakota, Utah, and Colorado during the first oil price boom following the 1973 Middle East War. Dick Davisson was an oil investor with White Weld & Co. and a mentor from my Boston analyst days at Colonial Management. He advised on our last meeting just before my moving to New York that I should be sure to look up Bill Barrett at Rainbow whenever I was visiting Denver. It was one of the best heads-ups that I have received in over forty years.

In 1977, Bill and his partners sold Rainbow Resources to the Williams Companies, and following the expiration of his noncompetition agreement, he began to build his new company, Barrett Resources, into what would arguably become the leading Denver-based public exploration independent during the last two decades of the twentieth century. In part, Bill's success turned on his uncommonly deep insight into the key geologic attributes of the Rocky Mountains, most notably, those of the Big Horn, Powder River, Green River, Wind River, Uintah, Williston, and Piceance basins along with the Overthrust Belt. Bill patiently and very attentively monitored activities throughout these vast geographical areas looking for geologic opportunities to capture scalable resource exploitation projects.

One spring day in the early 1980s, I joined Bill for lunch at the Denver Petroleum Club and noticed that he was unusually upbeat. On my inquiring why, he said he had just heard the great news that Mobil Oil had announced that morning that it was abandoning its efforts to commercialize the extraction of oil shale on Colorado's Western Slope near Grand Junction. He further explained that the major oil companies giving up on oil shale development would remove a key obstacle that had frustrated Barrett Resources' ability to develop a gigantic natural gas resource that lay in deeper zones beneath these same lands. Because surface mining was the then-existing technology for exploiting oil shale, codevelopment with natural gas wells was either awkward or impossible in much of the basin. However, as oil shale efforts across the industry shut down, the path was cleared for Barrett Resources to lead the way in assembling a large land position in this new play. Recalling Exxon's early June 1980 highly optimistic headline about its by-now-abandoned oil shale production goals (described in chapter 3), I began to realize that being in the right place at the right time (even if for an entirely different reason) had its merits on certain occasions.

Bill Barrett's plan to turn his initial idea into a fully validated and ultimately widely recognized success actually required years of experimentation and technical innovation to learn how best to drill, stimulate, and complete these wells. Progress was characterized by two steps forward and one step back as some new approaches worked and others did not. However, persistence and creative thinking about new techniques were generating net progress. The result was a learning curve that was scalably applicable to hundreds of thousands of acres with literally thousands of potential well sites. The overall effort required building a larger, highly professional organization capable of bringing a full range of engineering, geologic, and financial skill sets to focus on this large resource opportunity. Finally, it required a series of sound business judgments to navigate the challenges presented by low natural gas prices. These levels were periodically very depressed by an extended nationwide natural gas oversupply (the chronic "gas bubble") stretching for over a decade from the early to mid 1980s well into the 1990s.

In the mid-1990s, Bill displayed leadership skills that demonstrated how his executive judgment extended well beyond finding and developing new oil and gas fields. Plains Petroleum was an independent oil and gas producer with strong cash flow that became available for purchase. For a variety of reasons (largely owing to a challenging petroleum pricing environment), the company had been unsuccessful in finding a buyer. When Barrett Resources began to consider a transaction with Plains, my colleague Jon Hughes and I noted that there was considerable doubt within the company's board of directors and management about proceeding. There were debates about our presentations of the pros and cons of such a transaction. We observed some healthy skepticism about the specific terms available in the deal. Bill's quiet, even understated, leadership manner prevailed as he patiently allowed all objections and reservations to be voiced. He then persuaded, one by one, each of the directors and key members of management to embrace his strategic vision for consummating the deal.

Bill's management style reminded me of an old adage to the effect that we have each been given two ears and one mouth and that our listening and talking should be done in that proportion. Applying that approach enabled Bill to win the debate when it really mattered. His key point was that Barrett had developed sufficiently to need the financial heft and staying power that would come from a combination with Plains Petroleum.

Sometime during the negotiations, I noticed a picture that Bill had behind his desk. It was a depiction of a sled dog team moving across a snowy landscape. The accompanying inscription read, "If you are not the lead dog, the view never changes." This underscored in my mind that one should never take Bill's careful listening to be indicative of a lack of competitive drive. A year or so later, the wisdom of Bill's persistence in selling the Plains transaction to his board was vindicated when Barrett Resources experienced a serious well blowout at a delineation well in the high-pressure Cave Gulch gas field in Wyoming. Substantial corporate resources had to be diverted to extinguish the fire and regain control of the well. Barrett's ability to deal with this crisis in its drilling operations was much enhanced by its having completed the Plains Petroleum merger. After surviving that challenge, the company was able to continue developing its large Cave Gulch discovery. By the late 1990s, Bill had built the company into a leading oil and gas exploration company with a New York Stock Exchange listing.

In 2001, with Bill having retired and embarked with his wife, Louise, on a worldwide travel tour, Peter Dea was the Barrett Resources CEO who received the call from Shell explaining its unsolicited tender offer. Goldman Sachs had been previously retained by the Barrett board for defense issues. Nonetheless, upon hearing the Shell news and while working with Jon Hughes on the H&P/Key Production deal in Tulsa, I made a call that gave rise to an invitation to present a proposal for Petrie Parkman to become a co-advisor on defense. Given our firm's knowledge of the company and its assets, as well as of industry trends and issues more broadly, the board approved our engagement. Over the ensuing weeks, Petrie Parkman along with Goldman was able to assemble a data room presentation that outlined in detail the company's strategic positions and its impressive growth potential. Fortuitously, a timely regulatory decision by the Colorado State Department of Natural Resources allowing the company's request for an increased density of drilling in the Piceance basin enabled the Barrett team to identify about a half-trillion cubic feet of additional unbooked gas reserves as an incremental asset to be considered in valuing the company. Thus, Barrett Resources was able to develop multiple competing proposals that were superior to the Shell offer. Of these, the board elected to enter into a cash and stock merger with the Williams Companies.

With the interest of much larger companies in acquiring midcapitalization E&P enterprises now revealed, the next four or so years saw a series of consolidating transactions. Ultimately, there were twenty-two significant corporate acquisitions. Petrie Parkman's mergers and acquisitions team, led by Jon Hughes and consisting of Randy King, Mark Carmain, and Andy Rapp, was an advisor on fully one-half of these transactions. Among the more notable examples further confirming the interest of the larger-capitalization petroleum enterprises in getting back into the U.S. upstream sector were the following acquisitions: Mitchell Energy by Devon Energy, Westport Resources by Kerr-McGee, Western Gas Resources by Anadarko Petroleum, and Kerr-McGee by Anadarko. Another notable transaction was the acquisition of Magnum Hunter by Cimarex Energy. In their prior configurations, neither Key Production nor H&P's E&P affiliate could have managed such a strategic transaction. However, with H&P's and Key's total reserve base and their operations fully integrated to form Cimarex, this transaction was readily consummated. The spinoff of Key (originally valued at less than $30 million) from Apache has resulted in a company today with a capitalization of $6 billion, a fitting tribute to the late Mick Merelli and his successor management team now running the enterprise, as well as the combined board of Cimarex.

This wave of M&A activity was probably the most extensive consolidation of enterprises ever experienced in the U.S. independent exploration and production sector in less than a half decade. One of its most interesting aspects was the fact that it did not entail any discernible diminishment of the industry's ability to pursue new opportunities for resource development. Within a remarkably short time, virtually every management and technical team was made redundant. Thus freed up by the consolidation in ownership of the oil and gas reserves involving these transactions, all these teams were again back in business and well capitalized. This process was an expanded version of what had happened in the early 1990s following the sale of Tenneco's oil and gas operations. As Arthur Smith has detailed in *Something from Nothing,* it was a group of Tenneco petroleum professionals under the leadership of Joe Foster who built Newfield Exploration from scratch into a significant new upstream company. Other Tenneco alumni pursued similar paths. In the 2000s, this became the primary model for rebuilding the E&P sector. In most cases, there was also new financial support from a very strong group of private equity funds looking to position in the upstream sector. This phenomenon is a real-world example of how effective free-market capitalism seeks to embrace economically attractive opportunities in the energy sector.

Rising Fears of Global Peak Oil, 2000–2005

M. King Hubbert's disciples pick up where he left off.

M. King Hubbert died in 1989, but his pioneering analyses of peak oil events have continued to be pursued. Given the public attention generated by his original forecast about the timing of conventional U.S. peak oil, numerous Hubbert disciples took on the challenges of analyzing the prospect for and timing of peak oil production in other basins and countries around the world. Among others, these include Colin Campbell (an oil consultant), Ken Deffeyes (Princeton professor), Chris S. Skrebowski (editor of *Petroleum Review*), Richard Heinberg (peak oil author), Jean Laherrere (oil company geophysicist), David Goodstein (vice chancellor at Cal Tech), Thierry DesMarest (CEO of Total), and the late Matt Simmons (energy investment banker). Over the period 2000–2005, their collective work began to gain wider public attention. In the same period, U.K. and Norwegian North Sea oil production, in fact, did peak, essentially as predicted, after only three decades of aggressive exploration and development. This focus was further reinforced as the list of other countries experiencing declining production continued to lengthen. By 2005, well over one-half of the world's producing countries had entered what appeared to be irreversible natural declines in conventional oil output.

The onset of peak oil does not mean that we will be running out of oil in the short term, the medium term, or even the relatively longer term. In all likelihood, at least one-half century after the onset of global peak oil, there will still be significant production of oil. What peak oil does mean is that in the aggregate, a point will ultimately be reached at which new additions to the conventional global oil production base are insufficient to offset the annual decline in production from the planet's largest and more-mature existing fields. The timing of that event is probably being pushed out somewhat further than generally expected, given some of the same horizontal drilling and fracking technologies that are unlocking unconventional oil resources are now also beginning to be used in a variety of conventional oil fields, with encouraging performance. At a minimum, the widening application of such techniques seems likely to significantly moderate the rate of natural decline even in existing conventional oil and gas reservoirs.

Within the petroleum industry itself, the debate about global peak oil continues to be controversial, with both advocates of its inevitable occurrence and proponents of its ongoing deferability. The former observers base much of their case on the decided change in character, size, quality, and location of new conventional oil field discoveries that have been made around the globe over the past four or so decades. This pattern stands in sharp contrast to earlier periods such as the 1950s, 1960s, and 1970s in which there were many giant discoveries. In contrast, the optimistic observers argue that shortfalls in reserve additions are largely due to progressively more limited access to prospective acreage. They are convinced that new discoveries, and new extraction technologies, in particular, could unlock the much-expanded potential of unconventional oil resources. Such developments afford the possibility of significantly changing the outlook for future production. This is especially the case if restrictions on access to acreage are reduced or eliminated.

As often occurs in such debates involving global trends, there is some truth to both arguments. There has been an ongoing tendency toward much more restricted access to attractive acreage. This pattern began in the 1970s as the OPEC nations began to exert greater sovereignty over their oil development policy. During the course of that decade, the rules of the game (indeed, the actual reserve ownership) had changed dramatically for Western oil companies operating in the major exporting petroleum regions of the Middle East, Africa, South America, and Asia. This process is described in Anthony Sampson's book *The Seven Sisters,* published in 1975. The subsequent full playing out of geopolitical trends is even more comprehensively detailed in Daniel Yergin's Pulitzer Prize–winning volume, *The Prize: The Epic Search for Oil, Money and Power.* Both volumes have depicted critical historical events in the realignment of geopolitical and oil pricing power. With the benefit of today's hindsight, it is clear that one important effect of this power shift was to incentivize the world's major oil-exporting countries to push their resource planning and exploitation horizons out over multiple decades. In fact, in the case of the relatively few major exporting countries where the option was feasible, it involved taking as much as a fifty- to one-hundred-year time perspective. This entailed restricting access to new exploratory concessions and slowing development to ensure a more constrained balance of oil supply with demand.

Moreover, increasingly limited access to oil-prospective acreage has not been confined to areas involving the OPEC concessions. In the non-OPEC world, opposition from a variety of sources because of environmental issues and other technical concerns has reduced the scope of opportunity to explore for new fields and bring production to market on a timely basis. This has been especially the case in the frontier regions of the U.S. and Canadian Arctic as well as offshore both coasts of the United States. Even in the Gulf of Mexico, parts of the eastern gulf have been largely off-limits for many years. In recent years, this restrictiveness was beginning to change toward somewhat better access. However, there remained ongoing restrictions given the sensitivity of the Florida Panhandle and coastal residents and property owners. In the aftermath of the Macondo disaster of 2010, it is difficult to be very optimistic about how much additional constructive change there will be versus the current resistance to offshore development. The same attitude is becoming increasingly prevalent concerning portions of onshore federal lands under the control of the Bureau of Land Management and the Department of the Interior. Furthermore, the types of environmental opposition commonly demonstrated in the United States are showing up elsewhere around the world.

While improved access to prospective acreage might mitigate the onset of global peak oil, the longer-term magnitude of such benefits is not yet fully determinable. In any case, with regard to conventional oil-producing reservoirs, I believe that in view of the lead times involved to develop alternative energy supplies, we still need to consider the implications of an eventual (though not imminent) peaking in global conventional oil supplies. In the past, I have termed this "practical peak oil" and based this conclusion on several key assumptions and judgments. First, while access to prospective land may improve somewhat, it is unlikely to become a large enough factor to qualify as a game changer. Second, various analyses indicate that the embedded annual decline rate in the existing global producing base appears to be in the range of 3–5 percent per year. This means that at current production levels, new field output on the order of 2.6–4.4 million barrels per day is required annually simply to hold production flat. Achieving the high end of this range is a tall order. Such an achievement could be possible for several additional years, especially if the current optimism about new Iraqi production growth comes to pass and also if Saudi Arabia remains both politically stable and willing to expand its production at least moderately. Somewhat more optimistic views articulated by David Demshur, CEO of the reservoir optimization company Core Laboratories, and Daniel Yergin would suggest that 2.5–3.0 percent may prove to be the embedded global decline rate for the next decade or so. This could lower the annual production replacement requirement to a more manageable level, perhaps comfortably under 3 million barrels per day. In sum, these possibilities suggest the likelihood of a period of "plateau oil" before the actual onset of irreversible global production from conventional oil fields.

Global oil and related liquids production hit an interim peak of almost 87 million barrels per day in 2007 just prior to the global great recession triggered by the financial meltdown of 2008. Impaired economic growth worldwide resulted in a rather temporary decline of 3 million barrels per day in global oil demand. That setback has now been erased as the recovery in emerging markets has regained a degree of momentum.

In the view of some observers with whom I agree, whether the world production case can accommodate sustainable growth in daily conventional oil output much above 95–100 million barrels is unclear. Recent progress in Iraq, as well as planned expansions of oil production in the United Arab Emirates, Kuwait, and a few other exporting countries, may imply that such a view is somewhat too cautious. However, as news headlines periodically underscore, the nongeologic and other security-related obstacles to achieving the Iraqi promise remain formidable. Moreover, even if these planned new additions are brought on in a manner that essentially offsets embedded declines in the established producing base, global growth in demand for oil could still ultimately outrun additions to conventional global production capacity in the next decade. At that point, it appears likely that oil price levels will rise further, when needed to ration available supply as well as provide financial incentives for developing more-expensive unconventional and renewable energy sources. Finally, these are the conditions likely to enable sufficient development globally of unconventional oil resources to generate a most helpful extended period of "plateau oil" as an important mitigating factor in addressing the peaking of conventional global oil production.

As attention getting as M. King Hubbert's 1956 prediction proved to be, perhaps his even more noteworthy insight for those of us dealing with today's issues and trends is contained in his challenge implicit in the following statement: "Our ignorance is not so vast as our failure to use what we know." This Hubbert assertion of decades ago speaks volumes about today's tasks of dealing with pressing energy realities. In fact, much of what we have seen in terms of the industry's more recent efforts to evaluate and develop new resources over the past decade amounts to a strong embrace of this observation by Hubbert. We are getting better at using what we now know we know! Because of the emerging potential of unconventional resources, there are valid reasons to be optimistic. With what has evolved over the next several decades, the United States now has the ability to cope much more effectively with the challenges posed by international disruptions of oil supply and an eventual peaking of conventional oil production. With possible lags, the same may become true for many other oil-producing countries.

The United States Builds an LNG Infrastructure

They built it but so few came: surviving a profound change in the macro natural gas environment.

Another example of U.S. capitalistic enterprises moving to adjust to unfolding dynamic changes in the availability of hydrocarbon fuels involved the build-out of facilities to import large amounts of liquefied natural gas. At the close of the 1990s, concerns about North American natural gas supplies were once again on the rise. The long period of oversupply of gas had first appeared in the form of a mid-1980s "bubble" and ultimately became described as having elongated by the early 1990s into a "banana" shape. The surplus supply of natural gas ended about 1994. New discoveries of natural gas in conventional reservoirs were then beginning to decline once again. Progressively, ever-smaller accumulations for these types of fields were being drilled. In the offshore, quite surprisingly, the deepwater Gulf of Mexico was yielding much higher percentages of oil (and thus, proportionately less gas) than explorationists had expected. In the remote frontier regions of the Canadian and Alaskan Arctic, the long-proposed projects to deliver natural gas by pipelines from the McKenzie Delta and the Alaskan North Slope fields remained hopelessly stalled by a combination of limited project economics and complex political issues. By 1999–2000, the evolving common wisdom of many energy sector leaders was converging on the notion that a better answer to a threatened shortage of natural gas in the United States would be the importing of natural gas in the form of liquefied natural gas (LNG). It was expected that such supplies would come from known plentiful sources of previously "stranded" gas in West Africa, in the Middle East (especially Qatar), and throughout the Pacific Rim.

An early innovator acting on this insight was Cheniere Energy under the leadership of Charif Souki. This opportunity was, in effect, a real estate play for which "location, location, location" was the driving mantra. Several factors would determine the highest-value LNG locations for importing LNG. Cheniere secured options on four such sites in the United States. The selection criteria focused on four factors: (1) adequate depth of harbor (45 or more feet); (2) access to pipeline takeaway capacity; (3) a relative lack of NIMBY (an acronym for "not in my back yard") zoning obstacles; and (4) a worthwhile level of local demand for natural gas. Cheniere had earlier approached Petrie Parkman for advice and financing assistance on an unrelated project. In that case, my colleagues Mike Brock and Lon McCain had arranged a small private placement for Cheniere of which principals of our firm participated as investors. The company then focused on pursuing an LNG import strategy in the U.S. Gulf Coast. Having developed an appreciation for the economics of LNG beginning with my early-1970s visits to Marathon Oil, and following that up with insights gained while Petrie Parkman was working with Union Texas Petroleum, with its large LNG investment in Indonesia, we were very intrigued with the LNG opportunity. Cheniere subsequently adroitly pursued a variety of other transactions to build an impressive position in what were, for a time, four prospective LNG regasification (or "regas") facility sites on the U.S. Gulf Coast. By selling some of its sites, the company was able to concentrate its efforts on developing its Sabine Pass import terminal.

Not surprisingly, numerous other players began to consider LNG import and regasification sites. By 2004, there were more than three dozen identified and proposed LNG import sites throughout the United States. However, many of these fell by the wayside because of a lack of one or more of the previously mentioned factors determinant of ideal sites. In the end, only four new sites were actually built by various industry participants, along with some expansions at existing terminals. Altogether, these projects effectively doubled the number of LNG U.S. import terminals. Cheniere Energy's Sabine Pass facility was the first and largest ever new plant to come online. In total, these four sites increased the U.S. nameplate import capacity for LNG regasification to over 10 billion cubic feet per day, full operation of which could account for as much as 16 percent of U.S. natural gas supply.

Most remarkably, during the period in which these facilities were being completed, the environment that had fostered the LNG regas facility building boom changed radically. The presumed tightness bordering on shortage of domestic natural gas supply in the United States was averted. Actually (as detailed later), the gas supply outlook fully reversed into yet another extended period of surplus largely because of the development of new unconventional gas supplies from a series of extensive shale gas projects in Texas, Arkansas, Oklahoma, Louisiana, Colorado, and throughout Appalachia. Consequently, the need for U.S. imports of LNG was cut in half just as a quadrupling of import capacity became available.

In addition, Europe became the preferred delivery point for Atlantic basin LNG supplies because European consuming countries had greater needs for flexible, multiple sources of supply as an alternative to the increasingly questionable cost and availability of Russian gas exports. Finally, Japan was a very competitive bidder for LNG imports because of problems with maintaining operations of its aging nuclear power facilities. This was the case even prior to the country's disastrous tsunami-induced nuclear nightmare. It became even more evident thereafter. Accordingly, the advent of U.S. LNG infrastructure for imports proved to be a notable exception to the old expression "Build it and they will come." These projects were built largely on time and on budget, but to date, very few have come to utilize them. In fact, most of these import facilities are operating at only 10–25 percent of capacity. Furthermore, there are limited prospects for any material improvement in regasification utilization rates over the next decade or more.

Given the much-brightened outlook for growing amounts of unconventional U.S. shale gas production, many operators of these new facilities are now considering reconfiguring their plants from an import/regasification mode to a liquefaction/export mode. Because both processes utilize the same terminaling, dockage, and LNG tanks, a significant amount of the required physical plant is already in place. Thus, the lower additional expenditures and required construction time associated with converting these facilities to an export configuration ensure that they should be highly competitive versus undertaking a "new build" site. If executed as now envisioned, the transformation of some of these U.S. LNG sites will represent an astounding about-face. As such, it amounts to an excellent case study of the power of capitalistic private enterprises with reasonably free markets to adjust to a major (even overwhelming) shift in international trade balances and geopolitical conditions.

At the Sabine Pass facility, the shift from a regasification facility to an export capability should be relatively straightforward because some two billion cubic feet of its daily regas capacity is already available for immediate conversion. Thus, it enjoys good flexibility in replumbing its LNG facility to an export configuration. For some of the other facilities considering switching to LNG exports, there are contractual regasification commitments that will need to be restructured. Freeport LNG is an LNG regasification facility built in Freeport, Texas, by Michael Smith, the founder of Basin Exploration, an E&P company that he sold in the late 1990s. Its affiliate, Freeport LNG Expansion, has announced an agreement with a BP subsidiary to export about 625 million cubic feet per day to international markets. The company has indicated that it expects to begin shipments in 2017. Other similar proposals are believed to be in the offing from Cameron LNG, a unit of Sempra Energy, and from a joint venture involving Kinder Morgan and Shell Oil at the Alba Island regasification site in Georgia, as well as other projects at the Cove Point facility in Maryland and at Golden Pass, Texas.

The amount of U.S. natural gas export capacity to be created by converting regas plants will ultimately depend on how much shale gas production growth is developed in the coming years as well as the relative price between international markets for LNG and domestic U.S. natural gas. In addition, there may be political considerations relating to governmental approval of gas exports as well as continuing environmental opposition. Under most scenarios, it could approach 10–15 percent or possibly somewhat more of U.S. natural gas productive capacity. Having adequate gas export capacity to provide a "pressure relief valve" for the U.S. production network will be critical to dealing with seasonal and periodic cyclical gas supply surpluses. With such a capability in place, domestic gas producers will be assured of having a wider range of options for selling newly developed output, and price volatility should be reduced.

The need for natural gas export capability is further underscored by early signs of a potentially important success by Freeport-McMoRan involving exploration of its Davey Jones prospect in the deep shelf of the Gulf of Mexico. This new source of conventional gas resources is in very shallow water but involves an ultradeep horizon (about 30,000 feet) to penetrate below the salt trapping layer. The vision and leadership in this extraordinarily important venture has been provided by Jim Bob Moffet, a co-founder of McMoRan Exploration and chairman of Freeport-McMoRan. The company's partners include Energy XXI and Plains Exploration and Production. If the confirming well is successful, it could upgrade the natural gas prospectivity for a region encompassing some 30,000 square miles of the shallow-water Gulf Coast. By one estimate, this deep geologic trend could represent a potential additional gas resource of 150 trillion cubic feet. If it proves to be economic, this could be powerfully additive to the already bright outlook for U.S. natural gas resources.

From a policy standpoint, the Obama administration appears to be taking a constructive stance on LNG exports. The Department of Energy (DOE) has oversight responsibility for issuing export permits. It has already done this for the first two trains at Sabine Pass and the facility at Freeport LNG on the Houston Ship Channel. As various additional export proposals have surfaced, DOE has engaged a private consultant to analyze the macroeconomic impacts of growing U.S. LNG exports. That report was made public in December 2012 and found the impacts to be positive across all of a wide variety of the cases examined. In terms of process, the DOE will announce its decisions on permit applications on a case-by-case basis, and each applicant is also seeking permits from the Federal Energy Regulatory Commission (FERC) for the facilities involved. While there is still vocal opposition by some domestic manufacturers and petrochemical companies, at least regarding the magnitude of LNG exports to be allowed, the administration appears to be supportive of expanded LNG exports.

The Fourth Oil Spike

Commodities markets periodically render "votes of no confidence."

The cover story of the June 2004 edition of *National Geographic* magazine, addressing the issues of higher oil and gasoline prices, proclaimed, "Think gas is expensive now? Just wait, you've heard it before, but this time it's for real, we're at the beginning of the end of cheap oil." As 2004 unfolded, oil markets started to experience a combination of adverse trends that set in motion yet another upward spiral in oil prices. In contrast to the "third spike" that occurred following the 4-plus million barrels per day supply interruption as Iraq's 1990 invasion of Kuwait unfolded, this one had a very different cause. The growth in global oil demand accelerated as emerging markets in Asia (especially China), Eastern Europe, Russia, and Latin America all began to experience higher rates of economic growth. Advocates of peak oil were citing increasing signs that global oil production was struggling to keep pace with demand growth. Non-OPEC production was flattening out as North Sea production began to decline. Overall, the list of new fields with significant reserve volumes to be developed outside of OPEC was decidedly shortening. Within OPEC itself, there were also reasons to be concerned. The amount of surplus production in OPEC countries available to meet emergencies or shortfalls elsewhere in the world had dropped from 5 million barrels per day to about half that level. With daily world consumption running at 85 million barrels per day, a less than 4 percent cushion was very concerning to oil market observers. For perspective, an interruption on the scale of the Kuwaiti invasion by Iraq now would require more standby supply than the cushion that was then available could possibly provide.

Further compounding these concerns was evidence that the outlook for future production from Saudi Arabia was becoming more debatable. Matthew R. Simmons, a Houston-based investment banker, had published a book, *Twilight in the Desert,* in June 2005 asserting a series of questions about the ability of the Saudis to sustain, much less increase, their oil production. In a preview of his observations and findings, Matt visited Denver and shared his perspective with a group of petroleum industry leaders in the Rocky Mountain region. This occurred the year before the book was published. Accordingly, having already heard him speak to the issues, I read the book with great interest when it was published. I knew that it would be controversial, because Matt was directly challenging public assertions by some of the senior leadership of Saudi Aramco as to that company's ability to further increase its output. Matt's conclusions were based on an extensive review of technical papers presented to the Society of Petroleum Engineers (SPE) regarding operating problems and issues in the major Saudi producing fields. It was my impression that he had qualified each of his conclusions as he painstakingly reviewed the details in the SPE papers on a case-by-case basis.

Matt was clearly staking out a position in distinct contrast to statements that had been made by the Saudi oil minister, Ali al-Naimi. After reading the book, my sense was that Matt was on to something noteworthy regarding the limits of Saudi productive capacity and maybe even its ultimate recoverability of oil reserves. However, the many caveats cited by Matt were such that the timing of a rollover in Saudi output was not at all clear, or even as imminent as he apparently believed. Regarding this more cautious conclusion, Dr. Sadad al-Husseini, a former executive vice president and member of the Board of Directors of Saudi Aramco, had written an article published in *Oil and Gas Journal* on May 17, 2004, that provided reassurance about the near- and intermediate-term outlook for Saudi production. However, Dr. Al-Husseini qualified his support for increasing production to 15 million barrels per day by noting that maintaining such an elevated rate of production over the longer term would depend on the quality and quantity of reserve additions yet to be made. Dr. Al-Husseini has also made clear that he would prefer to see Saudi production capacity held at a more sustainable 12.5 million barrels per day. Matt was persistent in articulating his views. Because they stood in such contrast to the official Saudi Aramco version, they attracted considerable print and TV news coverage. Thus, while they now can be seen as premature with regard to the timing of expected problems, they clearly contributed to the evolving oil market's awareness and concerns about a global tightening of the oil supply/demand balance.

Against this backdrop, during the first half of the previous decade, other factors began to be additive to a growing, fundamentally bullish attitude about oil prices. A virtual civil war in Nigeria was periodically introducing production interruptions into the global oil supply equation. Sometimes these involved accidental explosions as destitute Nigerians attempted to tap into pipelines to steal oil and refined products. On other occasions, the problems stemmed from oil-worker strikes or notably effective military actions of local rebel leaders looking to establish leverage to renegotiate via a guerilla war a larger share of the production pie. These interruptions typically were temporary, but they periodically amounted to shortfalls of as much as several hundred thousand barrels per day. The attendant uncertainty as to timing meant that they created more than an inconvenience. This was especially true when they coincided with shortfalls from Venezuela, as President Hugo Chavez (now deceased) began imposing escalating demands and restrictions on oil companies operating in his country. His ultimate nationalization of selective oil companies further compounded the problem of declining Venezuelan output.

As the chart on pages xvi–xvii shows, after a series of oil price advances characterized by higher highs and higher lows in 2004 and 2005, starting in 2006 another dynamic began to seep into the oil pricing environment and market psychology of the period. In addition to the fundamental and geopolitical factors already identified, a somewhat new and much more intense trading activity began to take hold of oil markets. Oil began to be seen as a hedge against increasingly questionable moves by a spendthrift U.S. Congress that was showing signs of dysfunctionality. This view was further underscored by the looming reality that the second term of the George W. Bush presidency was approaching the traditional lame-duck phase of its last two years. Then the midterm elections of 2006 confirmed a change that resulted in Democratic control of both houses of Congress. Accordingly, investment decision makers became even more motivated to consider methods to protect portfolios against the consequences of adverse congressional actions, equally concerning congressional inaction, and a weakening U.S. dollar. Establishing a "long oil position" became a favored option of many understandably skeptical institutional investors.

In this case, oil futures markets became a vehicle for investors in effect to deliver a vote of no confidence in the direction the United States was heading. In sum, Democratic opposition to the sitting Republican president signaled the distinct possibility of no serious ability to address energy issues or other national priorities until the next presidential election. In a call to me, a former First Boston colleague, Craig Drill, has termed this the Washington inclination to "kick the oil barrel down the road." One sign of the increasing use of oil futures and energy securities as a portfolio hedge (aka, "vote of no confidence") was the large and rising volume of oil trading versus actual physical daily consumption for the oil commodity itself. For a variety of reasons, the daily volume of oil futures has long amounted to a multiple of the daily global consumption of oil. However, between 2004 and 2008, what had been a factor of three or four times consumption expanded periodically to six to eight times or sometimes even greater multiples. By late 2007 going into early 2008, "the tail began to wag the dog." More accurately, the role of the "green video screen" became prominent as oil prices began to contribute strongly to perceptions of an upward pull to the expected future value of the physical commodity.

Predictably, there reemerged congressional allegations of oil price manipulation, as has occurred on numerous occasions over the last several decades. However, whether the core issue was the decision of global institutional investors to use oil commodity investments as a store of value, somewhat akin to the traditional view of gold, is arguable. Thus, one can see rising oil prices as amounting to a vote of no (or at least declining) market confidence in congressional leadership. During the first half of 2008, the self-reinforcing nature of the oil price behavior described above became fully manifest. In May 2008, Michael Masters, a hedge fund manager of Masters Capital Management, testified before the U.S. Senate Permanent Subcommittee on Investigations on Homeland Security and Government Affairs that "what we are experiencing is a demand shock coming from a new category of participant in the commodity futures markets." He went on to explain that this new category of commodity players consists of a variety of institutional entities, including corporate and governmental pension funds, education endowments, and sovereign wealth funds, as well as other institutional investors. Masters labeled this group as "index speculators" and noted that they behave in a significantly different manner than traditional commodity speculators. He also observed that collectively they have grown to the point of accounting for more of the commodities futures contracts than any other market participant. In essence, Masters was suggesting that this new category of commodities player presented a risk of self-reinforcing expectations that could lead to a relatively new form of market behavior. The Masters testimony appears to have had a noteworthy influence on recent congressional thinking about commodities futures markets; many of the Dodd-Frank provisions relating to new regulation of commodities markets appear to have been motivated by the concerns that were raised in Masters' testimony. According to Bart Chilton, a commissioner with the U.S. Commodities Futures Trading Commission (CFTC), "Given what we've seen in the financial markets, it just seems to make sense that we would be inquisitive from an oversight and a policy perspective." In particular, the implementation of the Volker Rule, which limits principal transactions by market makers, is likely to trigger an undesirable contraction in commodity market liquidity for producers needing to hedge their production.

Meanwhile as May advanced to June in 2008, the upward swing in the oil markets took on a life of its own. The Saudis, as the de facto leaders of OPEC, became progressively more concerned about the effect on oil demand of rising prices as $100 per barrel oil moved to $110 per barrel, $120 per barrel, and beyond. The Saudis had seen this movie before. From prior cycles, especially that of the early 1980s, they knew that if the long-term demand elasticity for oil were to be triggered, altering consumption patterns, then the price cycle would most likely end poorly for oil producers—it did just that, as depicted on the right-hand third of the frontispiece.

I was a guest commentator on CNBC on July 2, 2008. Oil prices had retreated from a recent peak of $147.50 per barrel to about $125 per barrel. I expressed the view that a further downward test below $100 per barrel was likely, and the price could even reach $90 per barrel or perhaps $80 per barrel. Though I had acknowledged the prospect of more than a one-third drop in oil prices, I am still chagrined to note how much I underestimated the downside risk. Oil in the futures market quickly passed through the levels I identified and continued to fall precipitously until finally bottoming out ever so briefly at about $35 per barrel a few months later. With the benefit of hindsight, we now know that demand elasticities had, in fact, been triggered as the early stages of the global "Great Recession" began to take hold. Oil demand in developed countries (as defined by the Organisation for Economic Co-operation and Development [OECD]) declined for nine consecutive quarters from the second quarter of 2008 through the second quarter of 2010. In contrast, non-OECD demand held up fairly well in 2008. Nonetheless, it then actually declined slightly in the first half of 2009 before beginning a fairly robust upturn in the second half of 2009 and beyond.

What are the lessons learned? First, sharp, even radical price moves are often necessary to trigger changes in consumption patterns that impact oil demand. Second, once an oil price decline is set in motion, a price overshoot on the downside is often required to reverse the changes in demand elasticity triggered by the earlier spike. These were not new takeaways to be learned. In many respects, they were simply a painful reminder of the price elasticity testing events that both OPEC and oil consumers first experienced during the early to mid 1980s and relearned yet again in the late 1990s. Equally or probably more important, these are rules well worth remembering as we look to navigate the coming energy challenges over the next few decades.

Macondo Disaster

> In a technologically complex and "crowded" economic world, incredibly bad events can and periodically do happen; contingency planning is critical to ensure effective emergency responses.

The tragic Macondo disaster involving a blowout, explosion, and ultimately a complete destruction of the Deepwater Horizon rig in the Gulf of Mexico in April 2010 represented a grave setback for the companies directly involved (BP, Halliburton, and Transocean), the offshore oil industry generally, and to some degree the strategic energy positioning of the United States. The lives of eleven rig workers were lost, and almost five million barrels of oil were spilled. The environmental and economic disruptions to the Gulf Coast economy were extensive on a scale rivaled by only a few of that region's most extreme hurricanes.

The postcrisis analyses of what went wrong, who was to blame, and what steps need to be taken to prevent such occurrences in the future have yet to be fully completed, more than three years after the event. BP's internal analysis in September 2010 acknowledged some of the company's shortcomings but also placed shared responsibility for decisions and performance deficiencies on its service contractors, specifically, Halliburton's failed cement job and the Transocean rig crews' failure to recognize and deal with the presence and threat of escaping natural gas. Notwithstanding BP's settlement of some issues with the U.S. government and other harmed parties, the intense ongoing debate over allocation of blame for each of these parties is understandable. There is also the still-unknown full extent of future financial liabilities for these enterprises.

The subsequent January 2011 report by the seven-member independent commission established by President Obama cast a wider net in its assessment of the situation. The commission had no members with any direct operating experience in the technical or engineering aspects of offshore oil production. While the commission essentially confirmed BP's assertion that the blame for this disaster should be shared by each of the involved companies, it did place more of the onus on BP as the operator and largest working-interest owner in the project. Many of the critical shortcomings identified by the commission were attributed to BP.

In this regard, it is noteworthy that BP's practices and oversight were publicly questioned and implicitly or sometimes explicitly criticized by senior executives from several of its peer group companies, including Exxon, Chevron, Shell, and ConocoPhillips. This public criticism of a major oil company by its peers was one of the most unusual phenomena I have witnessed involving an oil industry environmental problem in over forty years. For example, it has no counterpart in prior crises such as Phillips Petroleum's 1977 Ekofisk blowout and oil spill in the Norwegian North Sea or Exxon's 1989 Valdez tanker spill in Alaska, two serious mishaps with widespread areas of environmental impact. J. A. Turley provides an in-depth discussion of the technical details about the Macondo blowout in his *The Simple Truth*.

Particularly important is that the presidential commission report went well beyond assessing blame in this particular case and concluded that there were industry-wide systemic problems that needed to be addressed to avoid the risk of a future offshore drilling disaster. Part of the problem was identified as inadequate government regulatory oversight. In addition, the industry's oil spill response plans and in-place spill mitigation and oil containment capabilities to deal with the spill were found to be inadequate. There was the assertion that the government's permitting approval procedures should be reassessed and substantially tightened. Finally, it was concluded that a liability limit of $75 million for offshore oil spills was an insufficient deterrent to preventing companies from tolerating sloppy and unsafe working procedures.

Even before the commission issued its findings and recommendations in January 2011, the Obama administration began to take remedial action regarding the government's regulatory and oversight role. Following the blowout, the administration quickly imposed an all-encompassing six-month drilling moratorium on offshore drilling in the Gulf of Mexico. This even included a forced cessation of activity for various other deepwater wells that had been drilling, despite the lack of specific evidence of problems or safety issues in the case of many other companies' operations. Moreover, asserting patterns of lax regulatory oversight and rubber-stamped approval of generic environmental impact studies as well as boilerplate contingency plans, the executive branch moved to disband its existing offshore permitting functions and created an entirely new Bureau of Ocean Energy Management (BOEM) under the leadership of Michael Bromwich, a high-profile career litigator. He was charged with completely overhauling the staffing, routing out any possible or perceived conflicts of interests, and focusing the BOEM on functioning as a much more intensive safety-driven and environmentally driven, interactive offshore regulatory agency.

Not surprisingly, these remedial actions, while generally considered necessary and desirable up to a point, have drawn criticism. In the aggregate, they were viewed by the industry as representing an example of regulatory overreach. Both congressional observers and industry participants publicly stepped up to express concerns about the interruption of permitting processes on an industry-wide basis. That action was viewed as having potentially very adverse economic implications for the Gulf Coast region in terms of employment trends. For the country as a whole, it was seen as unnecessarily exacerbating U.S. trade imbalances, economic growth, and employment trends. Senator Mary Landrieu, a Democrat, was particularly outspoken in her criticism of the Obama administration. She even used her ability to block Senate approval of a White House appointee to the Office of Management and Budget to express her extreme disappointment with the Department of the Interior's remedial actions. Regarding the impact on the U.S. trade imbalance, it was estimated that the cessation of offshore permitting would reduce domestic production by 150,000 barrels per day in twelve months and by 500,000 barrels per day over three years. At a $90 per barrel oil price, that production loss by itself would widen the U.S. deficit by almost $5 billion and over $16 billion, respectively, for the two periods cited. The *Oil and Gas Financial Journal* cited a recent analysis by RBN Energy that confirms that the decline in Gulf of Mexico production over the two-year period to June 2012 actually did reach 500,000 barrels per day. The article also observed that the latest Energy Information Administration (EIA) data indicate a recovery in output of about 200,000 barrels per day. This was primarily due to the startup of several new deepwater fields, as opposed to a recovery from declines in previously producing reservoirs.

Against this backdrop, what is to be learned and acted upon in the aftermath of the Macondo tragedy? First, a retooling of government oversight has already begun, and more requirements to navigate the permitting process are a foregone conclusion. Moreover, the risk of a longer-term pattern of excess regulation is not insignificant, because of the high economic cost in terms of lost jobs, a widening trade imbalance, and diminished future energy security. Accordingly, within the petroleum industry itself, there has been ongoing discussion about adopting a model patterned after the nuclear power industry's self-imposed and self-administered operating and safety standards utilized after the near-meltdown at Three Mile Island in the late 1970s. In that case, the nuclear industry explicitly recognized that all of its participants had a high degree of shared interest in the effective and safe operation of each and every U.S. nuclear reactor. In effect, a failure for one utility operator was a failure with severe adverse implications for all industry participants. Similarly, this is much the case in offshore oil drilling and even more so for deepwater offshore oil drilling. This factor helps explain the chorus of criticism by other industry leaders directed toward BP at the height of the crisis. Accordingly, the longer-term solution could involve having an active and aggressive peer group overview of operating and safety procedures, standards, and effective remedial actions. While this is a conceptually appealing model as this book goes to print, whether it will actually be embraced by the petroleum industry leadership remains doubtful, given the varying types of operations and size ranges of these companies.

In the meantime, there are at least two other private-sector initiatives to significantly improve the industry's ability to respond effectively to a Macondo-type blowout in the future. Nine of the largest companies have moved to put in place a more extensive strike force to deal with possible future spills. This joint venture is entitled the Subsea Well Response Project (SWRP) and is intended to provide for standby equipment to directly address the technical challenges encountered in the Macondo blowout. In addition, there is a second similar initiative to form the Marine Well Containment Company (MWCC). MWCC and Wood Group have combined efforts to form a U.S. offshore reserve response team. This team would consist of 100 personnel able to be quickly activated should MWCC's modular capture vessels be called upon to deal with a deepwater well control incident. As detailed in the *Financial Times*, MWCC's equipment consists of an expanded containment system (capable of handling up to 100,000 barrels per day). Its design calls for the separation of liquids from gas, then flaring the gas and storing the liquids pending their transfer to a shuttle tanker to move to shore. The facilities and personnel will be based in southern Louisiana, where they will be trained to operate and maintain the equipment. This effort envisions an ability to operate in water depths up to 10,000 feet. Both of the plans represent a commitment by the petroleum industry to build on what was learned in responding to the Macondo disaster. They are committed to have readily on hand state-of-the-art equipment and personnel to deal with such an incident.

Second, a review of the intermediate-term regulatory measures to tighten the permitting process indicates that the Obama administration was dedicated to imposing a myriad of new rules designed to address every shortcoming identified in the Macondo disaster. These include issues involving blowout preventer inspections and certifications; health, safety, and environmental protocols; and spill response capacity and capabilities. With these revisions now instituted, there are considerable industry concerns that there inevitably will be an unnecessarily burdensome stretch-out in the leasing and well permitting process. These new regulatory initiatives were acknowledged and considered justified by government officials as the necessary price to pay for enhanced safety and environmental protection.

Not everyone agreed with the latter proposition. Both the judicial and the legislative branches of the federal government have pushed back on the executive branch. In mid-February 2011, federal judge Martin Feldman of the U.S. District Court for Eastern Louisiana ordered the Obama administration to decide within thirty days whether to grant numerous permits for deepwater drilling projects. He was reported to term the government delays in action to be "increasingly inexcusable." This case involved a lawsuit filed by Ensco against then–Secretary of the Interior Kenneth Salazar.

Legislative actions after Macondo have been limited, given the split control of Congress with the Senate in the hands of Democrats and the House of Representatives ruled by Republicans. However, within the House, a bill was passed in early May 2011 with bipartisan support calling for the administration to accelerate oil lease sales both in the Gulf of Mexico and off the coast of Virginia. This was titled the "Restarting American Offshore Leasing Now Act." Given Democratic control of the Senate, a comparable bill is unlikely to advance in the Senate unless and until there is a change of leadership of that institution. Nevertheless, it is noteworthy that the House of Representatives vote was 266 "yes" to 149 "no" votes, clearly indicating meaningful support among Democrats for this Republican initiative. In mid-May 2012, U.S. Representative Doug Lamborn of Colorado Springs began pushing a bill that would require the Secretary of the Interior to approve or deny new leasing permits within thirty days. In addition, Representative Scott Tipton, another Western Slope Colorado Republican, signaled an intention to advance a bill prohibiting "de facto drilling moratoriums" because of bureaucratic rulemaking encumbrances. Given the current leadership of the Senate, however, there will most likely be no action on this idea.

One can hardly overstate the importance of getting the rulemaking, leasing, and well permitting processes right in terms of effective regulation to protect the environment while also achieving a workable permitting process for offshore drilling. This is especially true with regard to pursuing the remaining deepwater potential in the Gulf of Mexico. For perspective, it is important to note how the perception of the technical experts on the geologic merit of the deepwater gulf has evolved and matured as we have gained more-detailed knowledge of its subterranean character. In the 1980s and 1990s, as the petroleum industry was pushing progressively into deeper-water environs, there was the belief that these regions of the gulf would become progressively more natural gas prone than shallower waters.

Actually, just the opposite situation has frequently proved to be the case. The productive horizons being explored have demonstrated a decided tendency toward having a lower temperature window consistent with trapping relatively more oil with some associated natural gas, rather than exhibiting a tendency toward more dry gas accumulations. This was a very positive surprise, especially given the high U.S. oil import levels that then existed. Somewhat surprisingly, there have been a series of relatively large oil field discoveries ranging upward from 100 million barrels to at least one field complex in the billion barrel–plus range.

The USGS and other governmental sources have estimated that the Gulf of Mexico's remaining undiscovered resources in conventional reservoir traps are in excess of 40 billion oil equivalent barrels. If this potential can be economically realized, it would approximate the level associated with the U.K. North Sea. This is not a trivial resource option, even for an economy the size of the United States. Thus, this volume would be powerfully additive to the impressive oil resources now being developed in a variety of unconventional shale oil and other petroleum liquids projects in North Dakota, Texas, Oklahoma, Kansas, Colorado, Wyoming, and California.

Opponents of continued pursuit of oil projects often discount the importance of new supply development by comparing new resource additions to the prodigious level of U.S. consumption. Such projects are often characterized as representing only six months to a year or perhaps two years of U.S. consumption. For example, this line of argument has been used repeatedly (and misleadingly) to generate arguments designed to obstruct any progress on evaluating the hydrocarbon potential of the Arctic National Wildlife Refuge (ANWR). These criticisms fail to recognize that by adding consistently to domestic supply over many years, these types of projects can make a substantive difference to the security of supply in oil markets on the margin. Such predictably sustainable supply additions do help to mitigate U.S. oil import dependence and reduce the cumulative adverse consequences for U.S. trade imbalances.

Another critical dimension to getting the governmental response to the Macondo tragedy "right" is that there already are and will continue to be implications for U.S. exploration policies in other prospective offshore environs, especially in the deep shelf of the Gulf of Mexico, the Alaskan Arctic oceans and other Alaskan basins, and offshore both the East and the West Coast of the mainland United States. Furthermore, because the United States has traditionally set a standard that other nations are likely to adopt, there will undoubtedly be follow-on consequences for deepwater development off West Africa, Mexico, parts of South America, Australia, and much of Asia. Because oil markets are closely interconnected, each of these sources of supply holds significance for the petroleum supply/demand balance.

Shell Oil's efforts to explore in Alaska's Beaufort and Chukchi Seas north and west of the giant Prudhoe Bay field provide a useful case study of the American government's tendency to suboptimize by overregulating the pursuit of what could well be strategically important new oil resources. For example, about seven years ago, Shell acquired a large leasehold position in the federal waters off the Beaufort and Chukchi Seas. It has now spent over $5 billion preparing to drill in both of these areas. Shell's attempts to obtain a permit to drill have been repeatedly frustrated by a series of regulatory delaying actions and environmental lawsuits. In May 2011, many years into the process, Shell finally received a modest degree of encouragement that the Obama administration might be inclined to look favorably on a newly refiled permit application. However, the required approvals were not forthcoming on a timely basis, and in September 2012, Shell announced that it had run out of time to drill a well to a depth needed to test the prospective oil formation. Then delays due to sea ice incursions pushed the rig off location temporarily, and there was some damage to its Arctic Containment system that would require repair.

Accordingly, the company decided that it would drill as many "top holes" as possible and set surface casing for reentry by the drillship in the summer of 2013. These were drilled, but subsequently, the drill ship *Kulluk* ran aground off the Alaskan coast about one thousand miles away from the area of exploration focus. This event has prompted a new "Expedited Assessment of 2012 Arctic Operations" by the Department of Interior. According to the governor of Alaska, this study will determine whether Shell will be allowed to continue its activities later this summer. In January 2013, the EPA announced that it was citing Shell for violations of air pollution standards (running an incinerator longer than allowed). This appears to be another example of the environmental opposition's propensity to use all possible pretexts to hold up exploration activities. For this reason, the Shell permitting process was being closely monitored by other industry participants as an indicator of the U.S. government's sincerity regarding other new-frontier oil exploration initiatives. In April 2013, ConocoPhillips announced the indefinite postponement of its 2014 drilling plans for its leases in the Chukchi Sea pending clarification of federal regulations for Arctic Ocean drilling.

Regarding environmental activists' opposition to exploration, Alaska governor Sean Parnell has expressed his frustration with environmental activists who do not seem to credit Alaskans with having concerns about their state's environmental standards. He also noted that the profit incentive of oil producers includes a need to do operations correctly and safely. He concluded with the observation that federal actions on the Shell project would signal how serious the Obama administration is about its "all of the above approach" to energy development.

Key Lessons Learned

Mistakes and misjudgments can be even more instructive than successful decisions.

Distilled below are some of the key lessons learned and observations that I have personally come to internalize, drawing on more than forty-two years of developing investment research, advising corporate clients, and helping execute financings focused on the oil and gas business. My assessment in part II of the book of the workings of today's market forces and my reflections on America's future energy opportunities and challenges are through the lens of these time-tested perspectives.

Markets continuously endeavor to work, but in their own time and at their own pace. Repeatedly in the events previously described one can see that this adage is true of a wide variety of markets for oil tankers, crude oil, natural gas, oil field services, and energy-related investment securities (equities and bonds). There are often multiple factors affecting each of these markets. Therefore, it is important not to assume that there will always be a direct, obvious, or almost instantaneous correlation between a certain driver and a presumed market response. Inevitably there are strong corrective forces that kick in when markets drift (or spike) to excess. As numerous examples depict, the law of unintended consequences tends to be powerful and ever-present. Thus, regulatory measures designed to counter or control market moves often only further complicate or even obscure (at least for a while) market reactions to an unfolding geopolitical event or economic condition. More often than not, much as J. Howard Marshall counseled Secretary Schlesinger, it is much better to allow open markets to run their course rather than to impose arbitrary constraints that inevitably ignore or fail to adequately account for economic or commercial realities.

Flawed economic and policy incentives ultimately cause or exacerbate supply shortfalls (or sometimes even create undesirable surpluses). Examples of such incentives, signals, or unintentionally detrimental policy initiatives include arbitrary price regulation, mandated multiple price levels for chemically identical (or sometimes even just similar) commodities, governmentally mandated quotas such as for ethanol, or excessive emphasis on a green energy choice, to the detriment of more-cost-effective economic traditional sources that can still offer enhanced environmental benefits. Under such conditions, unnecessary shortages can occur despite the availability of ample raw energy resources. Most interestingly, as the U.S. and Soviet Russia respectively demonstrated in the 1970s and 1980s, both capitalistic and communist economies are capable of creating or exacerbating shortfalls amid an ample (or at least clearly adequate) supply of resources.

It is often darkest just before the dawn. In dealing with economic cycles, one should remember that doubling down on one's investment, or for that matter, on national energy policy commitments, at a near market bottom can be very rewarding, as long as it is a real bottom. Admittedly, identifying a true market bottom with certainty can often be difficult. However, in the energy sector when market values reach or exceed historic extremes, it is usually a lower-risk option to expect a reversion to the longer-term mean. Such is probably the case currently for natural gas (near its low-range value) and possibly for oil (which is nearer its high-side value).

Black swan events entail especially noteworthy risks. The implications of events broadly perceived as improbable are often far greater than many (indeed most) observers anticipate. This is because such outcomes, by definition, fly in the face of expectations for which the majority of market players typically have already taken steps to position themselves. For example, after the extended period of relatively safe offshore oil operations, the Macondo disaster involved the tragic loss of many lives and one of the most extensive marine oil-pollution events ever. Its financial consequences for the oil and service companies were very significant. Also noteworthy were the aggregate impacts on the broader U.S. Gulf Coast economy and the knock-on effects for U.S. offshore oil development, with further effects on the country's trade deficit. Given the industry's technical complexity, planning for perceived low-probability contingencies is extremely difficult. Nevertheless, risk mitigation can and must be improved to determine best safety and operating practices and to reduce the occurrence of highly disruptive events.

Powerful regenerative economic forces result from the application of well-incentivized capital focused on high-priority societal problems or needs. This benefit is easily underestimated, especially in highly charged or politicized times. The approximate threefold oil price recovery from a 2008 low of $35 per barrel has been a major driver of the recently transformed outlook for U.S. oil and gas production. Combined with the unusually high percentage of privately owned minerals to be developed in the Williston basin of North Dakota and eastern Montana, this price level has transformed the region into one having the brightest economic outlook in the country, with low-single-digit unemployment. These price signals also apply across the country. Improved access to undeveloped minerals, whether on federal, state, or private lands, affords the potential to generate roughly similar benefits elsewhere.

Old geopolitical grudges tend to reemerge, often at inopportune times and with adverse consequences. As history shows, these can lead to wars, capital destruction, and broad societal disruptions. There are historic precedents and continuing expectations as well as broader global needs for U.S. leadership in dealing with such events. Accordingly, it is important to maintain an awareness of possible trouble spots and a range of options to respond to them. The currently unfolding Middle East realignments involving Syria, Turkey, Saudi Arabia, Iraq, and Iran (among others) all pose numerous threats of triggering unresolved historic conflicts. Lessons learned from past successes and failures involving recent U.S. military commitments in Iraq and Afghanistan may be critical in forging future responses.

Periodic consolidation and reorganization (via mergers and sales) are integral to the evolving natural order of the petroleum-sector economy. Periodic consolidation and reorganization is an inherent aspect of the need for economic rationalization of market excesses both on the upside and on the downside. The three waves of petroleum-sector M&A consolidation occurring in the early to mid 1980s, in the late 1990s, and in the early to mid 2000s freed up a wide-ranging reservoir of human resources that unleashed powerful cycles of new entrepreneurial endeavors. These efforts were aimed at finding and developing new sources of hydrocarbon resources using new state-of-the-art technologies. Capital markets decision makers in both publicly traded companies and private equity platforms took the cue and responded impressively with ample financings for worthy projects.

Shifting global macro-economic drivers can overwhelm even a well-executed plan and thus necessitate midcourse adjustments in strategy for both corporate players and national entities. These can be likened to the shifts of tectonic plates in the earth's crust that give rise to momentous earthquakes. Each of the previous four decades has experienced at least one or two such broad-scale disruptive events in the energy sector. For example, the revolutionary changes in natural gas supply caused by advances in utilizing technologies to develop unconventional resources rendered invalid the need for natural gas import facilities in the United States. This resulted in a challenge for Cheniere Energy and many other energy firms to reconfigure their already built regas plants into export facilities and represents a major trade-deficit-reducing opportunity as the United States becomes an exporter.

Part II

THE ENERGY CHALLENGES (AND OPPORTUNITIES) THAT LIE AHEAD

Chapter 7

ALTERNATIVE ENERGY OPTIONS

The Foreseeable Future

"It's tough to make predictions, especially about the future."
—Yogi Berra

Consistently improving U.S. domestic energy production relative to imports over the balance of this decade is likely, along with reforms both in education and in immigration policies, to be among the factors most critical to American prospects for economic growth. In addition, it will enhance U.S. security interests in a world now experiencing major changes in the geo-strategic positioning of many of the more populous, growing economies, most notably those of China, India, and Russia. Recent advances in domestic oil production have already begun to favorably impact the historically chronic U.S. trade deficit. Further progress in reducing U.S. oil imports will be a plus in terms of both trade deficit improvement and mitigating U.S. reliance on foreign debt placements. More specifically, a much-lessened need for oil imports can reduce dependence on issuing growing debt to China and Middle East countries. This should enhance U.S. flexibility in dealing with the myriad of complex geopolitical issues involving countries that are not always naturally aligned with long-term American interests. Finally, the now-improved outlook for natural gas is in the early stages of restoring American competitiveness in key portions of the U.S. industrial economy.

The range of options for achieving greater domestic energy production is better today than I have observed at any time since the United States became a net oil importer. They include exciting potential for new oil and gas development utilizing various advanced technologies for tapping both conventional and unconventional sources of oil and natural gas, as well as a relatively small, though rapidly growing, contribution from alternative and renewable energy sources.

In addition, there are worthwhile gains to be made from much-improved efficiencies in the consumption of hydrocarbon fuels. The United States remains the largest per capita consumer of petroleum by a wide margin. However, likely progress to achieve higher miles-per-gallon standards for vehicles as well as similar efficiency improvements for appliances and other industrial and commercial energy functions afford opportunities to obtain the same or even more economic output for less energy input. Environmental advocates often assert that this is our largest single "new source" of energy to be tapped. While that conclusion is probably an overstatement, at least for the foreseeable future, there is little doubt that making strides with a variety of efficiency gains is an important energy strategy to be pursued aggressively.

The Problem of Scale

Renewable energy sources are growing rapidly,
but displacing fossil fuels remains a daunting task.

As encouraging as the supply outlook now appears versus that of even a decade or so ago, there are still formidable obstacles to overcome. The policy debates over tradeoffs between environmental goals and the energy needs of our society have been increasingly contentious and still present ongoing issues to reconcile. The simple fact that we are learning time and again is that there is "no environmental free lunch" when it comes to developing new energy sources on the scale needed in the United States. This is true globally for both the mature developed countries and the rapidly developing economies of the twenty-first century.

The energy density of petroleum (especially the liquid fossil fuels) is so extraordinary that it presents physical performance characteristics against which alternatives have great difficulty competing, in terms of both economics and actual physical accomplishments. A CSX Railroad television advertisement speaks volumes. It makes the point that the company can transport one ton (two thousand pounds) of freight almost five hundred miles on one gallon of diesel. A recent video documentary by Mark Mathis of West Wave

Films entitled *SpOILed—It's Time to Fill Up on Truth* made essentially the same case by pointing out that one barrel of oil can accomplish the work equivalent of five manual laborers working twelve hours per day for a whole year with no vacations or weekends off (that is, 365 days). Thus, the role of oil in advancing the quality of human lifestyles has been critical. No other type of energy source comes close to this intensity of output delivered in a relatively convenient package. Even more detail and discussion of this subject is contained in Robert Bryce's book *Power Hungry*.

The extraordinary progress of Western civilization over the past century and a half is explained to no small degree by the substitution of oil for coal, which had earlier replaced wood as a primary source of energy. These displacements resulted in unprecedented work productivity gains. Now that a wide range of high-population, less-developed countries are moving along similar paths involving more-energy-intensive drivers for their economic growth, the arithmetic of meeting those escalating needs, whether by fossil fuels or by renewable sources, remains very challenging. Notwithstanding the intrinsic appeal of the perceived "greener" categories of energy in terms of longer-term sustainability, it is sobering to recognize that each of these alternative sources has its own drawbacks in terms of higher costs, intermittency of supply, and other environmental impacts.

As we are now repeatedly discovering, even the green, renewable energy sources bring sometimes subtle but noteworthy environmental effects. In addition, despite impressive progress in reducing the costs of wind and solar, most such sources have yet to be proven broadly competitive with traditional energy sources without costly subsidies. In this context, the discussion that follows addresses a range of realistic options for improving the position of U.S. energy security and economic competiveness while giving increasing consideration to sensitive environmental issues.

To this point I have focused on relating lessons learned first as a petroleum investment analyst and subsequently as an investment banker in the oil and gas industry. In performing those functions, I have had to formulate and continuously update a point of view on the outlook for and economic viability of alternative energy sources. Accordingly, this chapter summarizes some thoughts about the realistic role of renewable energy sources as well as related limitations of and prospects for some of the major categories of alternative energy. Given my long involvement with the petroleum industry, some will deem these views as biased toward fossil fuels. However, I would be delighted to be proved too conservative (and even wrong) about these judgments regarding the potential as well as realistic time frames for green energy options to become a meaningful part of the U.S. (and ultimately the global) energy mix. In any case, many observers agree that given what we know today about green energy options, they are not and will not be a "magic bullet" for the foreseeable future. In *World Energy Outlook 2012*, the International Energy Agency (IEA) stated, "Despite the growth in low carbon sources of energy, fossil fuels remain dominant in the global energy mix." The IEA report further indicated that through 2035, the combined share of oil, gas, and coal in the energy mix can be expected to fall from 81 percent to 75 percent as renewables grow. In that light, the United States needs to pursue the best strategy for a mix of sources that includes both traditional and new unconventional carbon-based resources while still pursuing disciplined development of promising renewable technologies. The IEA report also acknowledged that a continued high growth rate for renewables will require ongoing large subsidies. Accordingly, in this era of constrained public budgets, judicious assessment of costs versus expected benefits will be critical.

Since 1971, there have been at least three occasions in which U.S. energy policy and oil supply crises have emphasized the development of alternative energy sources as a key to the transition away from reliance on fossil fuels. In the 1970s, after the first oil spike, there was an emphasis on using early-generation solar panels and wind turbines to generate electricity. At the time, the United States and other developed (European) countries were still generating large amounts of electricity with fuel oil. The more than tripling of oil prices in that early period followed by a second tripling in 1979–80 provided strong incentives to develop new and lower-cost methods for generating electricity, including some attempts using renewable sources. However, it rather quickly became evident that wind and solar technologies had inherent technical, conversion efficiency, and logistical limitations. Accordingly, in the 1970s and early to mid 1980s, the almost complete shift away from fuel oil for generating electricity in the United States led to increased utilization of a combination of coal, natural gas, and nuclear facilities. In turn, the near meltdown of a reactor at Three Mile Island in 1979, followed later by the full-blown failure of Russia's Chernobyl facility in 1986, ultimately curtailed reliance by U.S. utilities on nuclear development and further reinforced the use of expanded coal and natural gas to generate electricity. This first period of focus on developing alternative energy sources stimulated the beginning of a series of technological innovations leading to advances in solar and wind generation, especially as the need to improve upon the low efficiencies and attendant higher costs of these sources became evident.

The next period of rising interest in alternative energy sources occurred in the 1990s as the oversupply of oil and natural gas began to diminish, especially as the reliability of Middle East oil again came into question following Iraq's invasion of Kuwait and the ensuing First Gulf War. At this time, the role of wind turbines and solar generation had begun to expand again. The National Renewable Energy Laboratory (NREL) originally began operating as the Solar Energy Research Institute (SERI) on July 5, 1977. Subsequently, U.S. president George H. W. Bush elevated it to a national laboratory and changed its name to the National Renewable Energy Laboratory. This organization started to receive expanded funding after the first Gulf War in the early to mid 1990s for its research and development programs. In particular, this led to numerous advancements in wind turbine designs. The use of much longer turbine blades with improved gearing in new machines enabled them to operate with higher efficiencies over a much wider range of wind speeds and variability of wind intensity. These changes reduced somewhat, but by no means completely, the hazards for birds. This was especially important where generating facilities are located along paths for seasonal migrations. Similarly, in the area of solar research, NREL sponsored projects and other programs that evaluated a host of new materials and panel configurations to improve solar conversion efficiencies. In addition, initiatives were undertaken involving large-scale, centralized facilities for solar-to-steam generation of electricity. Each of these is an example of the benefits of effective governmental and scientific collaboration with private-sector enterprises to advance technically driven research projects.

The third phase of increased attention to alternative energy began with the upturn in funding for NREL projects by the George W. Bush administration following the 9/11 tragedy and extending beyond the Iraq War. The focus intensified as U.S. dependence on imported oil from Middle East sources continued to grow. This was especially true as the price of oil began to escalate sharply in the middle of the previous decade in the still-violent aftermath of the Iraq War. By 2009, the Obama administration was exhibiting strong interest in what it believed would be the transformative potential of green energy, and this became central to many of its energy-related decisions and policies during its first term. Funding initiatives continued to expand rapidly. Projects that had struggled to pass muster previously became reanointed as "shovel ready" following the 2008/2009 financial meltdown. Many of these became authorized under the $760 billion stimulus program with its now-recognized flaw of lacking adequate project accountability.

Wind, turbines, and solar energy projects continue to be among the leading candidates preferred by green energy advocates, for their characteristics of low or no greenhouse gas emissions and their inherent "renewability." Many of NREL's ongoing research efforts are directed toward improving the performance of energy generation by wind and solar devices. The focus is on conversion efficiency in generating power and reducing the cost per unit of electricity generated. Noteworthy gains have been made, and the rates of growth for installed renewable energy projects off the current small base in coming years are expected to be impressive. Nevertheless, there remain important cost limitations to the likely role these sources will play over at least the balance of this decade. Even the optimistic forecasters concede that the aggregate contribution of wind and solar over the next two decades is likely to account for only a single-digit percentage of total U.S. electrical generation. Furthermore, the capital that needs to be allocated to these renewable projects is still typically higher per unit of power generated than that required for a conventional energy source of comparable size. All in all, wider adoption of renewable energy projects remains hampered by much longer payback periods and higher amounts of front-end capital at the time of installation.

As we pursue electrical generation from wind and solar, we should consider other factors as well. Because both are highly variable generating sources over time, the need to store electricity when it is generated for use later when it is actually needed remains an ongoing significant challenge.

Many battery designs and other schemes to address this problem are being pursued, but solid breakthroughs so far are proving elusive. New battery designs often are based on the use of expensive new materials with limited availability. Accordingly, progress will likely be slow for some time.

For example, the issue of achieving adequate scale in addressing the storage of renewable energy has led the Brookings Institution and other macro-policy think tanks to contemplate the merits of a new "smart" electrical grid capable of delivering green-generated electricity to individual households where it can be selectively stored and, if needed elsewhere, periodically retrieved from the battery packs of hybrid electric or all-electric vehicles. A prototype of this ingenious idea for a storage system has been tested and in effect lab-bench validated. However, to expand it to the scale required for the vast U.S. economy of more than 300 million people involves a whole series of "chicken versus egg" obstacles. In other words, the degree of market acceptance required to achieve a critical mass of actually installed vehicle battery packs is a formidable task in itself. Beyond that, there is the required widespread installation of a brand new green grid for the necessary connectivity of this diversified storage system. Can the key capital-intensive components of such a complex be developed simultaneously to create this admittedly elegant vision of the future? A recent *New York Times* article detailing the travails of Fisker Automotive's efforts to build a competitive electric vehicle is instructive as well as sobering. The problem is further compounded when one contemplates the possible (and even likely) open-market challenges to electric and hybrid vehicles posed by natural gas as well as much more efficient traditional gasoline engine vehicles.

An additional somewhat related problem involves the need for standby generation that can be brought online quickly to replace wind turbines when the wind dies down and, equally important, can be readily shut down when the wind picks up. In today's world, natural gas is the main option for achieving this. As a rule of thumb, for every unit of wind-generated electricity, about one-half a unit of standby (that is, fully developed but usually nonproducing) natural gas generating capacity is estimated to be needed. In the current environment of modest economic growth and given the current level of excess natural gas production capacity, the ability to address this need for backup generating capacity is fairly easy to achieve. In future periods with the prospect of higher calls on the developed natural gas production base, however, the same may not be true.

Another way in which NREL has looked to improve the balance of the U.S. energy supply with demand has involved encouraging architects and builders to develop new energy-efficient residential, commercial, and industrial buildings. The Leadership in Energy and Environmental Design (LEED) program has fostered innovative designs using passive solar technologies, as well as heat pumps, improved natural lighting, and many other techniques, to reduce the level of consumption of fossil fuels and clean water supplies. The NREL headquarters in Golden, Colorado, is a case study in the aggressive use of such techniques. This newest building on NREL's almost-600-acre campus is considered the finest example of using today's state-of-the-art technology to achieve a net zero impact in terms of energy consumption. This is accomplished utilizing a combination of measures including geothermal heat pumps, passive solar options, active solar panels, highly effective insulation, and natural lighting techniques. The building has become a real-world example to architects and builders of what can be accomplished with creative forethought. My own experience with a LEED building was further shaped by what I observed while serving on the Board of Directors of the Gettysburg Foundation during the building of the new visitors center at the Gettysburg National Battlefield Park. Like NREL's showcase example, this building for the visitors center and museum utilizes an array of geothermal heat pumps, passive solar, and other advanced lighting techniques to achieve worthwhile energy savings.

Effective design of energy-efficient buildings will become increasingly important over the balance of this century. However, in the United States, a large amount of infrastructure is still in place and is likely to remain in use for a long time. It was built in previous eras of low energy prices, presumed relatively low future energy prices, and much lower environmental sensitivity. The time frame to accumulate enough energy savings to be nationally significant realistically needs to be measured in multiple decades as opposed to a few years or even the one or two decades often articulated or at least implied by the many advocates of alternative and renewable energy sources. In this regard, it is noteworthy that emerging economies have a decided competitive advantage in that if they build with energy efficiency in mind, their "new build" projects would, by definition, account for a much higher percentage of energy consumption, given their relatively small currently installed base, than is the case for the United States and other more mature Western economies.

Beyond the traditional types of alternative energy projects, NREL and others have also focused on developing a variety of biofuel projects to manufacture ethanol from corn, agricultural refuse, and other cellulosic biomass sources. In particular, the introduction of tax credits for ethanol helped advance many such projects. Initially, ethanol was being used as a supplement (or blending agent) to gasoline to boost the octane level of gasoline. Because of its carcinogenic attributes, methyl tertiary butyl ether (MTBE) became outlawed as a competitive alternative to ethanol for raising octane levels. The latter issue and MTBE's demonstrated tendency to leak from tanks and to seep into aquifers, thereby threatening drinking water supplies, posed a serious threat to human health. Consequently, ethanol was left as the primary option for refiners to meet required octane levels. Thus, ethanol became a growing part of the U.S. gasoline supply, approaching 10 percent.

With the tendency of some in positions of leadership to assume that more is almost always better than less (at least until it becomes obvious that is not the case), various individual states in the United States began to weigh in with mandates to move the percentage of ethanol up from 10 percent to levels ranging from 20 to 30 percent. A few years ago, I personally witnessed this type of thought process in action while speaking at a conference of Governors of Western States. Similarly, in June 2012, the Environmental Protection Agency (EPA) approved the sale of gasoline with an ethanol content of 15 percent. An even more extreme approach would involve the idea of vehicles equipped to run on E-85 (a gasoline/ethanol mix, respectively, at 15%/85%). As it turns out, configuring automobile engines to run on E-85 fuel is the relatively easy part of the overall task. In fact, many cars now being manufactured already have the capability to switch to that fuel configuration.

However, it has become apparent that other obstacles to implementing a shift to higher percentages of ethanol in the fuel mix are formidable. Ethanol is difficult to transport. Thus, it effectively requires a completely separate (indeed, duplicative) fuel distribution infrastructure. Moving it by pipeline is generally not feasible, because of its affinity for water. Furthermore, trucking ethanol fuel is much more expensive. Therefore, ethanol's optimal role seems likely and more appropriately to remain limited to being an additive to gasoline as an octane enhancer at about 10 percent of the total gasoline fuel mix in transportation. Furthermore, as boat owners have become well aware, increasing the percentage representation of ethanol in the gasoline distribution system can also entail negative consequences. Higher ethanol content has the potential to put at risk parts of some marine engines and associated fuel lines.

In addition, more-detailed analysis of the presumed benefits of ethanol use as a fuel has raised various doubts as to whether there is any actual net gain in energy efficiency or environmental benefits from substitution of ethanol for gasoline. This stems from the lower energy content per unit of ethanol as a fuel, as well as a more complete appreciation of ethanol's less than ideal emission attributes. Finally, there is rising awareness in the United States that the adverse effects of growing corn as a feedstock for ethanol involve other detrimental environmental consequences, such as an enlarging offshore "dead zone" for Gulf Coast shrimp caused by much-expanded fertilizer supplies draining into the Gulf via the Mississippi River. Similarly, in Europe there are indications that recent more-comprehensive analyses of the environmental impact of biofuels are raising additional questions about previously presumed benefits. As these issues are becoming better appreciated, the drawbacks are beginning to be weighed more carefully. There is also the fact that at its recent peak, about 40 percent of U.S. corn production was being used to generate the current level of ethanol production. Thus, there is the argument that a rethink of the size and appropriate role for ethanol in the U.S. transportation fuel mix versus food needs and related cost impacts is in order.

Advocates of ethanol tend to discount the issue of corn as the ethanol feedstock by citing the hoped-for potential for generating ethanol from cellulosic materials. In the long run, whether the United States can compete effectively in growing cellulosic materials versus more favorably geographically and climatically positioned countries such as Brazil, Colombia, and others in Latin America remains to be seen. More recently, there is mounting evidence that ethanol projects are becoming seriously economically challenged by near-term oversupply. In November 2012, the New Energy Corp. ethanol plant in South Bend, Indiana, filed for bankruptcy. The *Wall Street Journal* reported that the winning bid of $2.5 million was based on its "scrap value." In another case of a distressed asset sale, Bionol Clearfield was reported to have sold for $9 million versus a reported original cost of $270 million. After three decades of almost uninterrupted growth (only one down year out of thirty-three), ethanol production declined for a second time in 2012. This may be the marketplace underscoring to policy makers that ethanol is best utilized as an oxygenate but not as a primary transportation fuel.

There are two final points being made against further growth in ethanol as a percentage of the gasoline pool. First, given the need for large amounts of water in ethanol production, the recent drought has raised doubts about its utility as a gasoline additive versus food needs. Second, recent industry statements indicate that rich ethanol credits are driving up the prices of transportation fuels.

By way of summary, there is no denying that the pursuit of alternative energy sources has broad-based consumer appeal and merit. There have been many technological advances in projects that are promising of future benefits. Nevertheless, from the standpoint of policy implementation, there are noteworthy limitations to what we can accomplish in a realistic time frame. Lawrence Mone, president of the Manhattan Institute, has observed that based on the latest U.S. Energy Administration's annual energy outlook, solar and wind are unlikely to exceed 3 percent of total U.S. energy needs by 2035 unless there are technological breakthroughs. While this view may be too pessimistic, even a tripling or quadrupling of the role of renewables will fall short of significantly impacting the need for cleaner fossil fuels to accommodate economic growth.

The IEA is considerably more optimistic than the Manhattan Institute about the overall ability of the OECD nations to increase the representation of renewable sources in the mix by 2035. I suspect that this is more likely a stated goal than an analytically derived prediction. That said, it must be acknowledged that ongoing research and development programs in the governmental and private sector alike have made substantial progress in lowering the costs of renewables. There is merit in continuing such research efforts, but many valid questions have emerged about the appropriateness and cost-effectiveness of large government subsidies to accelerate introduction of yet to be commercially proven renewable energy projects. In any case, recent results suggest that the timeframe for renewables to become a significant source needs to be carefully judged.

Climate Change Issues

Polarized positions almost always guarantee inaction.

Few issues are more polarizing than the matter of how to deal with the implications of today's heightened fears of global climate change. A few years ago during a meeting with a Silicon Valley client, I made an observation about former vice president Al Gore's book and the subsequent movie to the effect that "the inconvenient truth about *The Inconvenient Truth* was that it is not all true!" The individual with whom I tried to share that thought literally came out of his chair to express his outrage with my controversial assertion. Not surprisingly, our discussion quickly degenerated into talking past one another. From what I have observed in the public discourse, my California experience is being replicated repeatedly in many venues throughout our society. Unfortunately, the role of objective analysis in assessing options and alternatives for dealing with the issue of global warming and its implications for climate change has been deeply subordinated by a fairly widespread set of programs to assert a high likelihood of severe consequences if there is anything less than an extreme and rapid makeover of our current energy systems to produce and deliver sustainable, renewable energy to reduce the role of fossil fuels (in other words, the imperative to go fully green).

Nevertheless, there is the sobering argument that by the time we have greater certainty about our assessment of the role of human actions in creating the global warming challenges, it could be too late to take appropriate and effective remedial actions. Accordingly, now is the prudent time to take steps to reduce the amount of greenhouse gas emissions caused by human activity in the twenty-first century. Interestingly, one of the more effective measures already under way is the use of much more natural gas (instead of coal) to generate electric power. By various estimates, this could reduce carbon dioxide by 50 percent per unit of electricity generated. An analysis in

U.S. ELECTRICAL GENERATION SWITCH AND EFFECT ON EMISSIONS

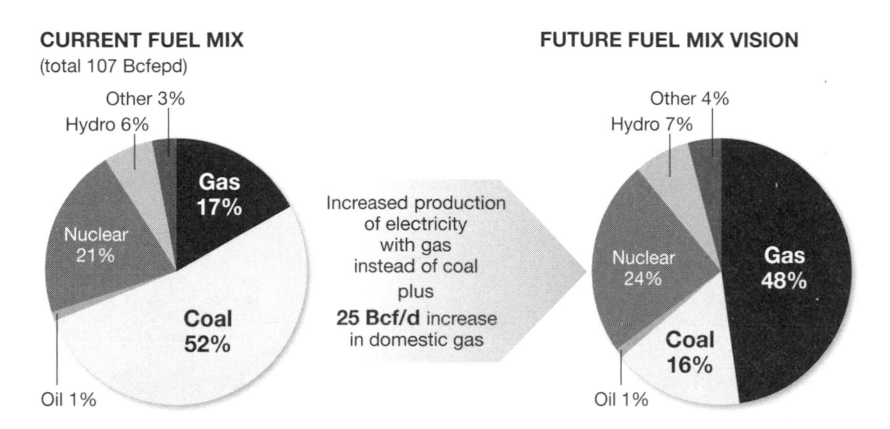

EFFECT ON EMISSIONS

Metric tons

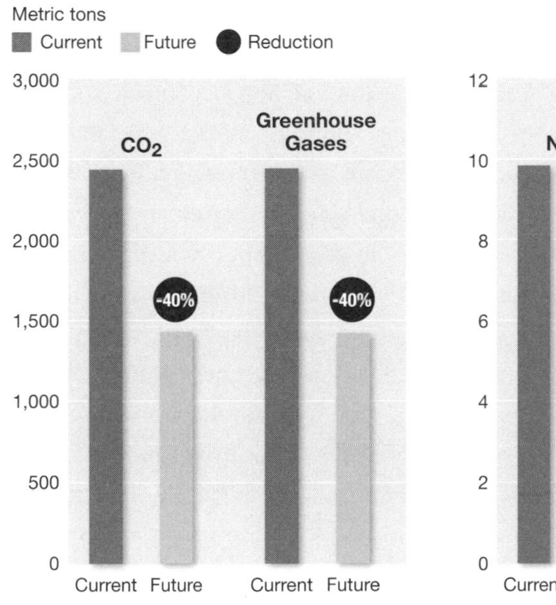

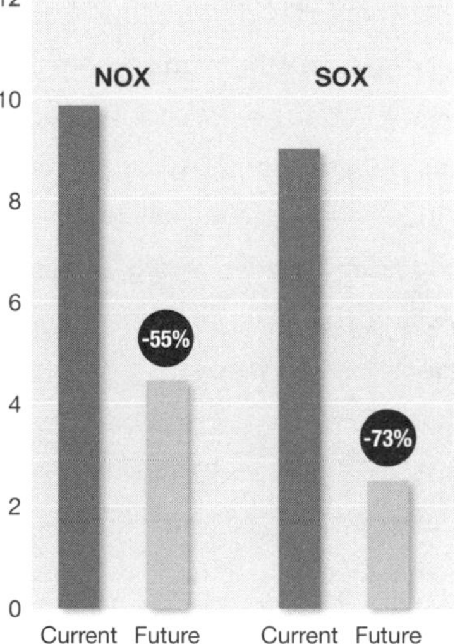

SOURCE: EnCana projections 2008.

2008 by EnCana showed the emissions reductions possible if 25 billion cubic feet of newly developed natural gas were to displace coal as the primary fuel for generating electricity in the United States. In addition to an estimated 40 percent drop in carbon dioxide levels, natural gas would yield even larger percentage reductions in the levels of both nitrogen oxides (NO_x) and sulfur oxides (SO_x) as well. As outlined below, progress along these lines has already been made, but more is possible.

With the makeup of the U.S. Congress in terms of the number of coal-producing states, natural gas seems unlikely to achieve its full potential benefits in the near (or even intermediate) term. However, even a partial success with the gas substitution for coal burning seems likely to yield more progress over the next decade or two than any other feasible option identified to date. It would buy the much-needed time and economic flexibility both to assess further climate change models and to make progress on the development of green or renewable alternative energy sources on a scale that is relevant without inviting destabilizing economic impacts. Furthermore, it provides the time and incentive to develop appropriate (and hopefully economic) clean coal technologies. Recent trends are encouraging. Coal, which has historically accounted for over one-half of U.S. electricity generation, slipped to a 43 percent share in 2011, and in 2012 again declined sharply to under 40 percent, largely because of natural gas substitution and the retirement of old coal-fired plants. According to a *Financial Times* report, coal generation of electricity in the United States has recently declined by 19 percent, while gas generation increased by 38 percent for the same period. Interestingly, the United States is now ahead of what might have been its CO_2 reduction goal had it signed the Kyoto Protocol. Ironically, many of the actual signers of that agreement are behind schedule in meeting their CO_2 reduction goals.

This dramatic transformation reflects an important changing mindset on the part of utility company executives. They are capitalizing on the relatively low price of natural gas, which is in a price zone that is cost-competitive on a BTU basis with coal. This is especially so when the positive environmental attributes are taken into account. EPA rule-making decisions are encouraging the utility industry to focus on decreasing carbon and other emissions. In December 2009, the EPA declared that greenhouse gases are a danger to public health and thereby set the stage for proposing new regulations for electric power plants, smelters, refineries, and other emitters of greenhouse gases (including carbon dioxide). Over the following two years, coal and other industry opposition to the new rules took the form of increased lobbying followed by litigation questioning the EPA's rule-making authority. In June 2012, a federal appeals court upheld the EPA's moves to restrict greenhouse gas emissions. That decision ratified the EPA's 2009 findings; unless it is appealed and then reversed by the Supreme Court (which seems unlikely), it puts in place a powerful incentive for utilities to continue to move toward burning less coal and more natural gas in their generation of electricity. In fact, recent EPA deliberations indicate that expanded curbs on new coal plants are likely.

This conclusion is further reinforced when one considers the lack of other viable and scalable options to generate growing amounts of electricity needed to fuel economic expansion. While coal supplies remain plentiful, the environmental constraints of burning coal and the apparent absence to date of cost-effective technology to clean up coal emissions suggest that this is unlikely to be a source of growing electrical production in the United States. In fact, the EPA has acknowledged that many older plants unable to comply will ultimately need to be shut down. The output of those plants is estimated to be less than one-half of 1 percent of U.S. generating capacity, but depending on future rule making, that could expand over time. In any case, actual growth of coal-generated electricity seems doubtful unless or until new and economically viable technological advances are made in cleaning up coal emissions.

Likewise, the tepid outlook for future growth of nuclear power generation may be increasingly doubtful despite its intrinsic superiority to produce electric power generation without greenhouse gas (GHG) emissions. Other than one new facility planned in Georgia and some ongoing small expansions at existing facilities, there is little in the construction pipeline. In recent years, limited consideration has been given by regulators to the idea of streamlining the permit approval process by utilizing uniform reactor designs as is done in France. That idea has not seemed to gain much traction to date. In fact, rather than pursuing a single common reactor design along the lines of France, the United States seems incapable of reducing its number of designs to less than eight or nine. In addition, a June 2012 U.S. Court of Appeals ruling stated that the U.S. Nuclear Regulatory Commission (NRC) can no longer accept assurances that a permanent nuclear waste repository is coming. This decision reflects the fact that, motivated in part by opposition from senators and representatives in the affected states, the Obama administration has decided to cease work on the Yucca Mountain storage facility in Nevada. In effect, this action may amount to a unilateral abrogation of contracts with numerous electric utilities throughout the United States. Given the absence of a viable spent fuel disposal alternative at this point, the NRC has announced that it will stop issuing licenses for nuclear plants until the waste storage issue is resolved. As pointed out in a *Wall Street Journal* article on this decision, "The standstill in finding a permanent American nuclear waste dump could undermine the expansion of nuclear power, which is already facing a challenge from cheaper natural gas."

All in all, it is becoming apparent that there are currently no easy or credible answers to the question of what "follows" oil and natural gas. The scale necessary to achieve worthwhile reductions in the use of these fossil fuel–based energy sources is daunting. Studies to date acknowledge a lack of economic competitiveness (without costly subsidies) for many of the renewable options. For example, the U.S. wind production credit is estimated to have amounted to $6 billion over the period 2009–2013. An additional stimulus package subsidy for the same period amounted to another $8 billion. Perhaps as important, these subsidies introduced unintended consequences in the form of adverse impacts on otherwise attractive and competitive energy sources such as natural gas. When one begins to appreciate the magnitude of development needed for renewable sources to account for a meaningful double-digit percentage of the total energy supply pie, it becomes clear that the likely transition to a greener mix should probably be measured in terms of three to five decades or possibly even longer. Accordingly, what has become evident is the wisdom of pursuing relatively cleaner fossil fuel development involving natural gas to buy the extensive time that is still likely needed for technical advances to bridge to a greener energy future.

Chapter 8

UNCONVENTIONAL FOSSIL FUELS

Natural Gas: The Key Bridge to the Future

The chemistry of carbon matters.

In the second half of 2005, Petrie Parkman was hired as an advisor on the sale of a private company's holdings in the Barnett Formation, a hydrocarbon-bearing shale underlying a large area in the vicinity of Fort Worth, Texas. After considerable preparation, the sale process was initiated in early 2006 and resulted in two transactions with an aggregate value of more than $2.6 billion that closed in the second quarter of that year. This highly successful outcome represented a monetization of an asset base of Chief Oil & Gas, a company started from scratch in the mid-1990s by Trevor Rees-Jones, a private oil and gas entrepreneur based in Dallas, Texas. Over time, Rees-Jones' strategy evolved into the idea of producing natural gas from a formation previously long considered too tight (that is, characterized by a low permeability to the flow of gases or liquids) to yield economic volumes of hydrocarbons. These 2006 transactions almost immediately stimulated broader interest both across a wide segment of the petroleum industry and in the stock market. The large potential for the relatively new category of "unconventional" gas reserves was exciting and attention getting.

Over the past several decades, evolutionary innovations in drilling methodologies and well completion techniques began to demonstrate the feasibility of much-expanded production of natural gas from tight rock formations (those exhibiting very low permeability). As the price of methane improved following the end of the natural gas bubble in the mid-1990s, the necessary financial incentives finally became available to utilize the much more advanced rigs that could efficiently drill horizontal wells over lengths

ILLUSTRATIVE HORIZONTAL WELL

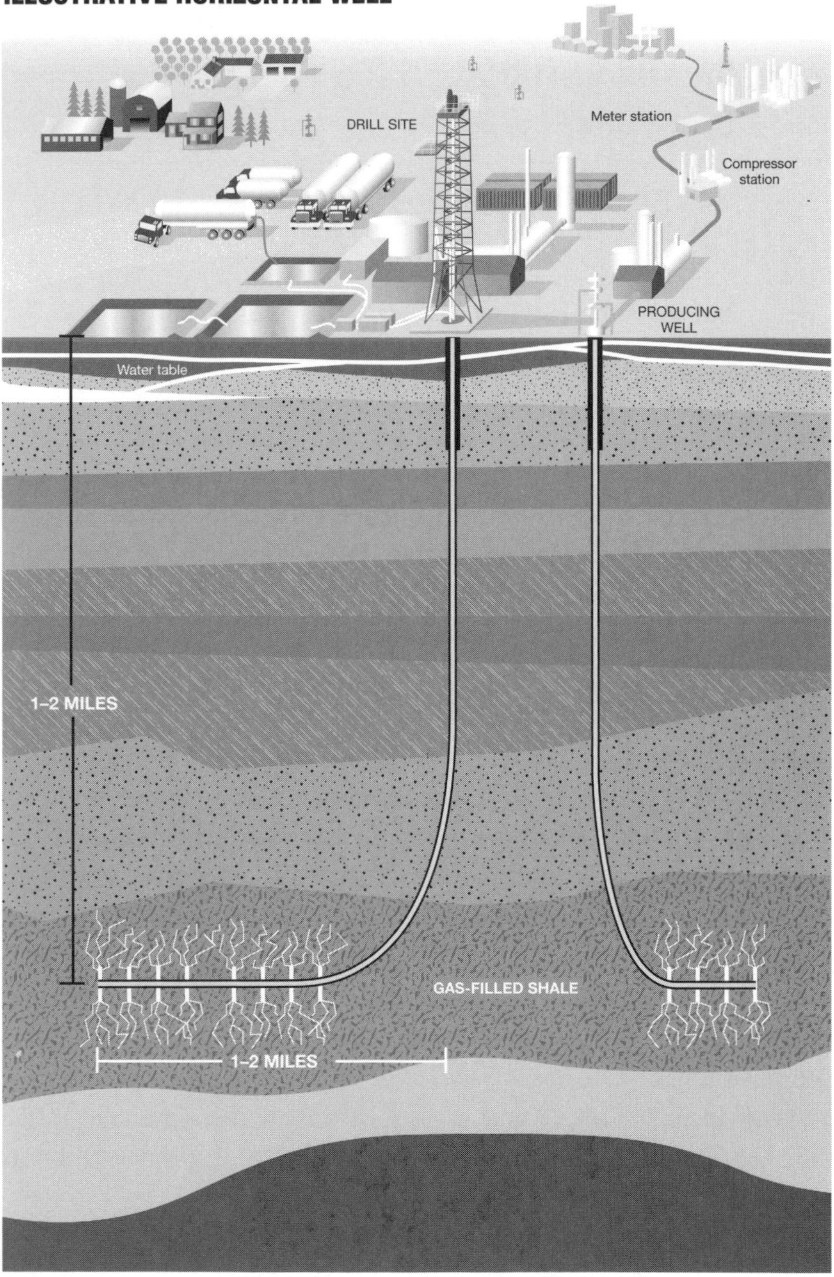

DRILL SITE

Meter station

Compressor station

PRODUCING WELL

Water table

1–2 MILES

GAS-FILLED SHALE

1–2 MILES

extending up to a mile or more into a prospective horizon. Thus, these wells have opened access to a much greater cross section of hydrocarbon-bearing formations. When coupled with a series of improvements in techniques to stimulate natural gas production by effectively breaking up the rocks containing the hydrocarbons, these new approaches became an economical way to recover hydrocarbon reserves in formations that had previously not even been considered prospective. This process is accomplished by injecting fluids at multiple intervals along the horizontal leg of the well under high pressure (hydraulic fracking). At first, the fluids used involved various gels, but in due course it became evident that slickwater fracs were much more effective. With horizontal wells now extending to lengths of as much as one to two miles, the number of locations (or stages) being fracked can actually range from twenty to forty (and sometimes more).

Many of these advances were pioneered by George Mitchell (now deceased), an independent oilman I first met in Boston in 1971 when he was launching an initial public offering of his company, Mitchell Energy. Over about a seventeen-year period beginning in the mid-1980s, he tenaciously pursued the challenge of finding an effective formula for unlocking resources from the Barnett Shale. This effort culminated in what he later jokingly referred to as his "overnight success." As the progress of Mitchell Energy and Chief Oil & Gas became more widely recognized by their competitors (including, among others, Devon Energy, Southwestern Energy, Petrohawk, EOG Resources, and Chesapeake Energy), it set off an ever-widening search for similar opportunities. First this occurred with the broadening of activity to develop the Barnett Shale in North Texas. Then the search expanded to shales with similar attributes in other areas. In March 2007, less than a year after the aforementioned large private sale of Barnett Shale properties, more than a dozen other new shale plays in hydrocarbon-bearing basins throughout the country were identified in Hart Energy's *E&P* magazine as being actively pursued by a variety of industry participants.

Several hundreds of miles to the northeast in Arkansas, Southwestern Energy's CEO, Harold Korell, recognized that the Fayetteville Shale, which was the same geologic age as the Barnett, also had intriguing potential to be economic. The company acquired over 800,000 acres covering the Fayetteville Shale. Aided in part by the trial-and-error learning curve experiences of

Mitchell Energy and Chief Oil & Gas, Southwestern Energy was ultimately transformed over the course of a decade from a $250 million company in the late 1990s to more than a $12 billion enterprise, reflecting its much-enhanced prospects for major production growth. In fact, it was George Mitchell's sending of a copy of Dan Steward's study of the Barnett Shale development that enabled Southwestern to compress its learning curve in developing the Fayetteville Shale. On reading the book, Korell realized that Southwestern was making the same early-on mistakes as Mitchell. Thereafter, he bought numerous copies of Steward's book and directed his key employees to not replicate previously determined unsuccessful practices such as using gels and fracking too close to natural fractures.

In 2005, Floyd Wilson, the CEO of Petrohawk Energy, decided to radically change his corporate strategy by focusing on opportunities in unconventional gas resource development. Wilson had previously built and sold two very successful independent producing companies (Hugoton Energy and 3-Tech). Petrohawk engaged Petrie Parkman to sell most of its holdings in conventional fields and used the proceeds to reposition itself to pursue unconventional shale exposures successively in the Fayetteville, Haynesville, Marcellus, and Eagle Ford Formations. Upon its later sale to the Australian mining company BHP, early Petrohawk shareholders were rewarded with a very significant expansion in the company's market capitalization.

Over the past half decade or so, many other company managements have followed up with projects in these and other prospective shale formations. For example, Devon Energy, Cimarex Energy, Continental Resources, and others have aggressively pursued development of the Cana-Woodford Shale in Oklahoma. Numerous other endeavors by a variety of operators have now validated much of the recent shale development activity across North America. It is important to note that not all shales are equally productive or comparably economic. Each shale formation can provide somewhat different results. This can be a function of its total carbon content and thickness, the brittleness of the shale, and its proximity to key consuming markets or to the available pipeline grid, as well as its degree of natural gas liquids (NGL) content (that is, liquid fuels including propane, ethane, butane, and pentane). The formation thickness may help determine the amount of hydrocarbons trapped in the rock, but not everyone agrees that that is a key

POTENTIAL GAS PRODUCTION RATE THAT COULD BE DELIVERED BY THE MAJOR U.S. SHALE PLAYS

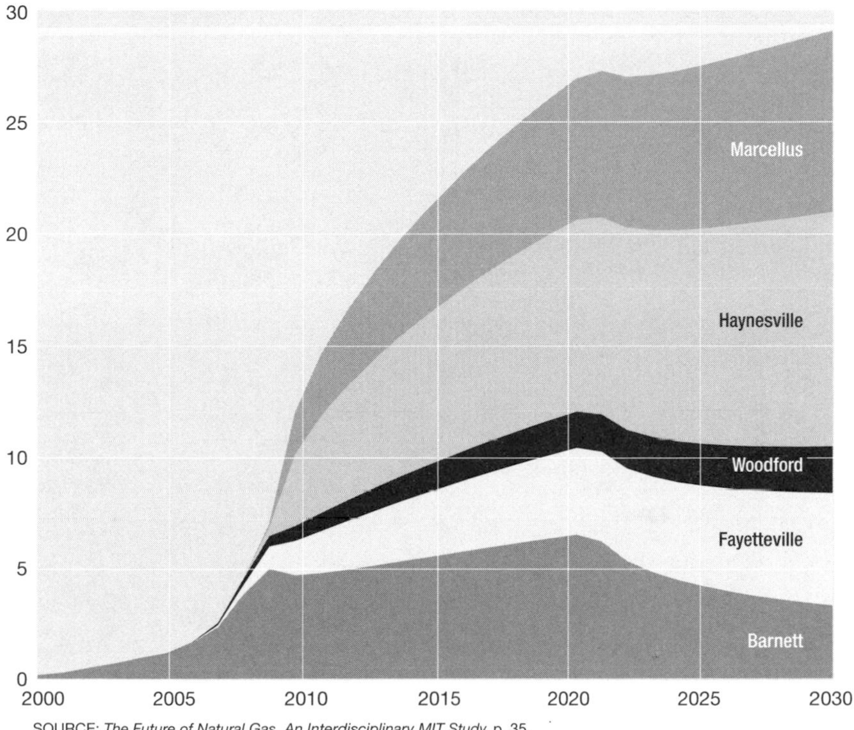

SOURCE: *The Future of Natural Gas, An Interdisciplinary MIT Study,* p. 35.

driver of economic success. The brittleness determines how well the formation will respond to stimulation by fracking. If instead of being adequately brittle, the rock is relatively soft (like peanut butter), the ability to break it up with fracking is unlikely to release hydrocarbons. Access to key markets tends to govern the level of price to be received for production as well as the speed and ease of hooking up new wells to the pipeline grid. Finally, NGL content enriches the heating value of the gas and can improve its ultimate selling price on a BTU basis. There is an impressive potential for production growth over this and the next decade for five of the more active shale gas development plays now under way. *The Future of Natural Gas,* an in-depth interdisciplinary study undertaken by Massachusetts Institute of Technology and published in 2010, has documented the "game changer" potential of shale gas development.

VALIDATED SOURCES OF UNCONVENTIONAL SHALE GAS AND OIL SHALE PRODUCTION

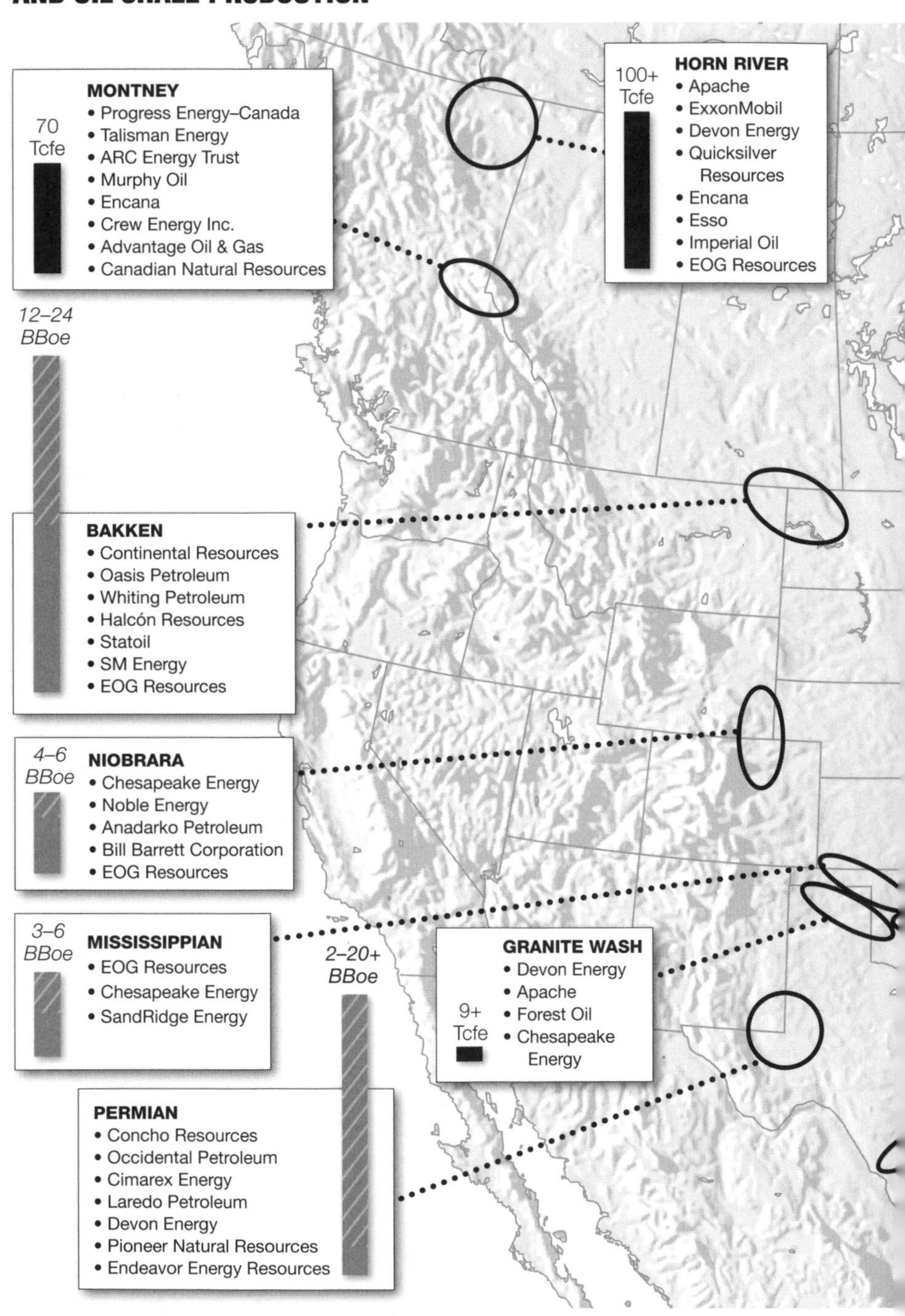

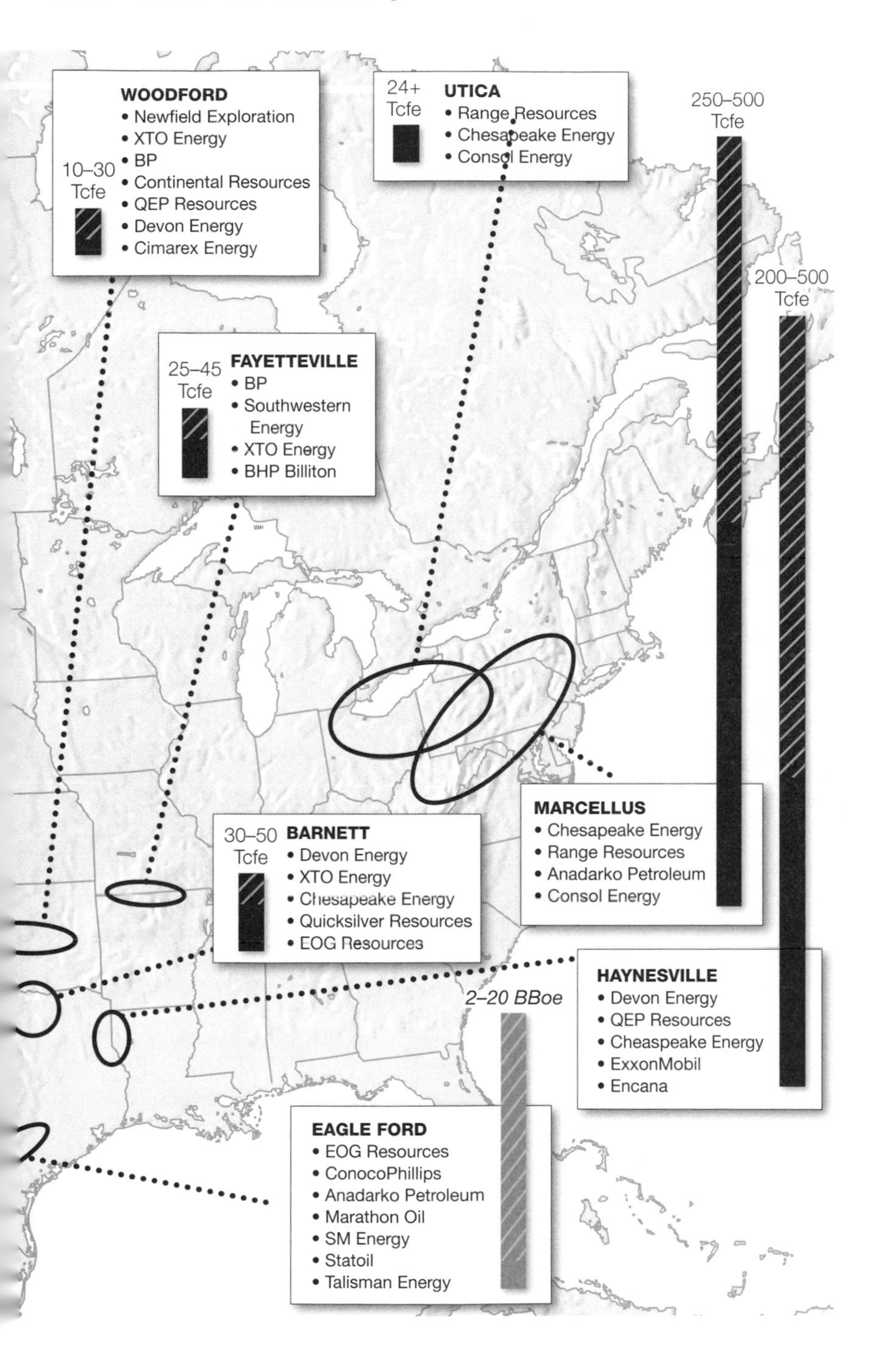

Oil
Gas
Predicted range
WOODFORD
• Newfield Exploration
• XTO Energy
• BP
• Continental Resources
• QEP Resources
• Devon Energy
• Cimarex Energy
10–30 Tcfe
24+ Tcfe
UTICA
• Range Resources
• Chesapeake Energy
• Consol Energy
250–500 Tcfe
200–500 Tcfe
25–45 Tcfe
FAYETTEVILLE
• BP
• Southwestern Energy
• XTO Energy
• BHP Billiton
MARCELLUS
• Chesapeake Energy
• Range Resources
• Anadarko Petroleum
• Consol Energy
30–50 Tcfe
BARNETT
• Devon Energy
• XTO Energy
• Chesapeake Energy
• Quicksilver Resources
• EOG Resources
HAYNESVILLE
• Devon Energy
• QEP Resources
• Cheaspeake Energy
• ExxonMobil
• Encana
2–20 BBoe
EAGLE FORD
• EOG Resources
• ConocoPhillips
• Anadarko Petroleum
• Marathon Oil
• SM Energy
• Statoil
• Talisman Energy

In recent years, the pursuit of unconventional gas resources has dramatically upgraded the potential for this fuel to play a key role in the future U.S. energy mix. For the first three decades of my career (1970s, 1980s, and 1990s), the outlook for natural gas could be described as having, on average, a horizon of forward visibility of about ten years. The production shortfalls in the second half of the 1970s were so severe that there was the fear that the United States might effectively run out of natural gas within a decade. Instead, exploration activities during the period from the early 1970s to about 2000 were such that the ten-year outlook tended to roll forward year after year. Put another way, current production was about a tenth of our known recoverable reserves, and thus we were finding about as much as we were producing.

This situation began to change radically when the idea of developing shales became feasible. Shales are often the source rock that generates hydrocarbons, only a portion of which would ultimately migrate to be trapped in much more highly permeable (and thus conventionally productive) natural gas reservoirs. Thus, being able to economically unlock hydrocarbons directly from the shales themselves or from immediately adjacent low-permeability (tight) rocks has significantly increased the size of the natural gas resource base that can be developed. Major industry players now estimate that our remaining gas resources could fuel current needs for as much as one hundred years. Because there still remain many uncertainties about how the future will unfold, it is probably more appropriate to view the forward visibility of the gas development option as ranging somewhere between fifty and one-hundred-plus years. In any case, the notion of a five- to tenfold (or more) expansion in forward visibility is a remarkable improvement for policy makers looking to address how secure future energy supplies can be brought to market in support of economic growth.

It has become evident that the unconventional gas revolution is not going to be confined to the United States. Projects involving the Horn River, Montney, and numerous other shales in Canada are already being pursued. One plan envisions developing gas for LNG exports to Pacific Rim markets via a liquefaction facility planned to be built at Kitimat in British Columbia. There are also projects under way designed to evaluate and validate shale gas accumulations in some of the eastern Canadian provinces. In addition, shales in the United Kingdom, North Africa, Poland, Hungary, and other Eastern European countries as well as Mexico, China, India, and potentially Australia are all now being considered for their unconventional gas potential. All in all, the odds favor a lengthening list of prospective shale gas resources in international locations as this decade unfolds. However, given the nature of capital to pursue first its highest-return projects, the pace of many international ventures is likely to lag that of the United States.

To gain technical and operating knowledge about exploiting these potential resources, many foreign companies have entered into more than a dozen joint ventures aimed at exploiting U.S. shale gas projects. Their motivation is to accelerate their own ability to develop shale gas resources in basins outside the United States. Such projects could be particularly helpful and environmentally beneficial in China, and possibly India, where coal still dominates the current energy mix. In September 2012, China announced that it is inviting foreign operators to participate in a series of joint ventures to develop shale gas resources on some twenty identified blocks in southern China. The proposed format contemplates Chinese-controlled foreign joint ventures with competitive bidding. The challenges include very little existing infrastructure, a lack of downhole geologic knowledge, and a shortage of technical expertise. These issues notwithstanding, the resource potential and local market needs are as big as (or potentially even bigger than) in the United States. However, because of these and other issues, the pace of shale gas development in China and potentially other countries is likely to be slower than in the United States. Nevertheless, as recent history suggests, the Chinese ability to set goals and achieve them often tends to be underestimated. Their additional incentive is that as the recent U.S. experience demonstrates, increased utilization of natural gas provides by far the most effective way to reduce both toxic and greenhouse gas emissions.

Because of the shale gas revolution, there is now reason to believe that the natural gas resource potential for the United States could well exceed 2,000 trillion cubic feet. To realize the full potential of this massive resource, a key determinant of the rate of future activity will be the development of new demand for natural gas as a substitute for more-carbon-intensive fuels. Growing the demand for natural gas could make a difference in three important sectors: the petrochemical industry, the electric power generation industry, and the transportation sector. Domestic petrochemical producers already are and should continue to be among the most immediate beneficiaries of what is unfolding. The now-ample and relatively assured future supply of natural gas has rejuvenated the U.S. chemical processing industry in a manner unimaginable even just a half decade ago. Most notably, the use of natural gas and especially its associated natural gas liquids (NGLs) in new facilities to produce ethylene, one of the primary products from which many plastics are made, is likely to remain highly competitive in the global marketplace for many years to come. This contrasts sharply with the situation during the previous decade when I personally witnessed reports of Chinese interest in buying uneconomical ethylene facilities on the U.S. Gulf Coast. The plan was to disassemble them, load the hardware on barges, and ship it to China for reinstallation on the Chinese coast, where the low cost of labor was then a competitive advantage. In contrast, U.S. ethylene-based petrochemical plants today have a decided and likely longer-term sustainable cost advantage over European and other foreign ethylene plants, which use much higher-cost oil-based feedstock to produce ethylene and its derivatives.

The second area of expanded use of natural gas involves the generation of electricity via gas-fired turbines. This may well be the option with the largest benefit, for several reasons. First, coal, which as recently as five years ago accounted for over one-half of U.S. power generation, remains a large emitter of many forms of pollution, including greenhouse gases, particulates, and mercury. Greater use of natural gas would virtually eliminate particulates while reducing carbon dioxide, sulfur oxides, and nitrogen oxides emissions. It would also go a long way toward reducing the mercury emissions. Because of these pluses, utility managements are already moving to embrace expanded utilization of natural gas to generate electricity. Furthermore, the U.S. nuclear power system currently involves over one hundred facilities, many of which are over thirty years old. As these plants come up for recertification, many are likely to be decommissioned because of age and noncompliance with safety requirements, necessitating the development of replacement capacity. Third and finally, developing new gas-fired generating capacity usually involves the lowest front-end capital outlays as well as the shortest lead times to project start-up. Accordingly, natural gas has the potential to claim an increasing share of the power generation market and could eventually almost double its market share. Considering the strength of the coal industry advocates in Washington, achieving only a portion of this potential seems a more prudent expectation for the medium term, but under many scenarios, market share gains by natural gas seem likely to continue over the decade.

The third area for expanding natural gas use involves the transportation sector. Here the challenges remain formidable, but the longer-term promise of transformation is more than intriguing. Natural gas is used today in about 12 million vehicles worldwide. Of these, less than 200,000 are in the United States. In contrast, Pakistan, Iran, Argentina, Brazil, and India (each with over a million such vehicles) have led the way to date in using natural gas to power vehicles. In aggregate, these five countries account for three-quarters of the global gas vehicle total. Thus, the possibility of greater and accelerating penetration of the domestic market seems reasonable and may even be inevitable. However, achieving a timely build-out of infrastructure to support greater use of gas in vehicles also poses another "chicken versus egg" (that is, which needs to come first) conundrum regarding infrastructure. To overcome this, Boone Pickens has proposed a plan to focus the attention on utilizing natural gas in the form of LNG in the long-haul truck market. He points out that if 60 percent of the 8 million eighteen-wheeler truck market were to be converted to natural gas, the reduction in U.S. oil imports could be on the order of about 2.5 million barrels per day. This would represent about a $90 billion annual reduction in the current U.S. trade deficit.

To be sure, such noteworthy benefits are not likely to be achieved quickly, but as Pickens has observed, they will not be achieved at all unless we start moving toward the goal. There are numerous obstacles to overcome in implementing the Pickens Plan. First, because the American Petroleum Institute indicates that natural gas vehicles are about $70,000 more expensive to build than a conventional diesel truck, an incentive program for buyers of such vehicles would help accelerate the building of the required fleet. Second, the systems for delivering natural gas fuel to such vehicles would need to be put in place. Today, such refueling facilities exist only in limited localities, largely for localized fleet vehicle use. The ability to establish such an infrastructure along strategic points in the interstate highway system seems feasible. However, the challenge is to coordinate the timing of installing such a system so that it coincides with the availability of a critical mass of natural gas–powered vehicles, thus addressing the classic "chicken versus egg" problem. Clearly, both efforts would need to progress simultaneously: the promise of one is required to stimulate the other. The potential scope of Pickens' concept is both vast and somewhat daunting, but so are the large economic benefits to be gleaned from a successful execution of the concept. For ground transportation in North America, it might ultimately begin to rival the economic multiplier effect once realized from President Dwight D. Eisenhower's embrace of the interstate highway construction program.

Beyond the Pickens Plan, progress is being made to expand the use of natural gas in other vehicle applications throughout the United States. Until recently, perceptions of the potential future growth of the non-18-wheeler natural gas vehicle market in the United States have not been optimistic, but there are now developing reasons to begin to think more positively. The most logical market to convert initially involves fleet vehicle applications in localized community delivery functions. These can take advantage of economies-of-scale leverage from centralized refueling facilities. That these types of utilization of natural gas in fleet vehicles have not grown more quickly is somewhat surprising, because the difference in prices for oil versus natural gas would seem compelling.

The benefits for vehicles that have traditionally relied on diesel fuel are especially noteworthy because of the high cost of producing ultra-low-sulfur diesel fuel from oil, as now required by the United States. Natural gas affords intriguing potential as an option to displace diesel fuel in certain fleet vehicles. Of the U.S. vehicle population, total energy usage is the equivalent of almost 22 trillion cubic feet. About 25 percent of these vehicles utilize diesel. Thus, a potential to substitute 40 percent of natural gas for diesel would amount to about 2 trillion cubic feet, or about an 8 percent increment to the current level of U.S. demand for natural gas. The benefits of such a substitution are several. First, elimination of imported oil equivalent to almost 1.0 million barrels per day with what is likely to be a sustainable lower-cost fuel could be feasible. Second, there is a reduction in greenhouse gases. According to NGVAmerica, the nonprofit public education arm of the American Gas Association, these reductions would amount to 22 percent and 29 per-

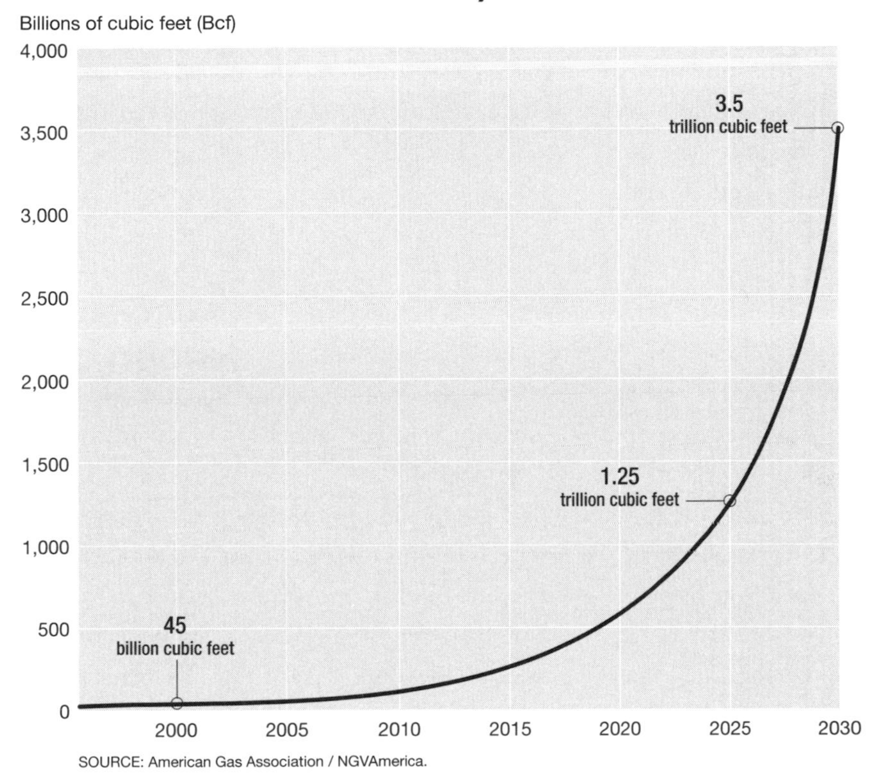

cent, respectively, in the case of diesel vehicles and gasoline vehicles. In addition, noteworthy reductions in other urban pollution, including ozone levels, sulfur oxides, nitrogen oxides, and particulates could be achieved. Thus, while there is no panacea for replacing liquid petroleum fuels quickly, the selective use of natural gas in U.S. vehicles affords the potential to mitigate diesel fuel consumption with a noteworthy positive environmental impact.

In addition, the natural gas vehicles represent a relatively attractive option because the required technology is already state of the art. Put another way, no new technical breakthroughs are needed in engine technology. Thus, as the work of NGVAmerica has suggested, while numerous options exist (including natural gas, ethanol, methanol, propane gasoline/diesel hybrids, plug-in hybrids, and natural gas hybrids), issues of scalable supplies, optimal effectiveness, and practicality suggest that natural gas affords the potential to become one of our most effective choices.

Furthermore, NGVAmerica remains very optimistic about the rate of penetration of natural gas vehicles into both the local fleet market and the individual owner vehicle market during the coming decade. This view is based in part on a "bubbling-bottoms-up" process of innovation to give both corporate and individual consumers a much greater array of attractive choices in terms of efficient gas-fueled engines. Many of these vehicle types are detailed on NGVAmerica's website (www.ngvamerica.org). Another proposed path for private individuals is to fuel their natural gas–powered vehicles at home. This involves a low-pressure refueling system that can operate when the car is parked in the household garage for the night. However, there are issues of possible methane leakage that need to be resolved for this option to become more widely adopted.

In sum, the role of natural gas in the transportation sector remains to be determined, but the prospects are promising. Future developments will depend on favorable governmental policy and private-sector innovations. The pace of acceptance will probably lag the absolute rate of penetration by natural gas into the power generation sector. Nevertheless, the benefits to be gleaned are clearly substantial and, for certain applications, compelling. For example, recent moves by FedEx and UPS to introduce natural gas vehicles into their fleets are encouraging. An expected 9 percent penetration of natural gas vehicles into the global vehicle fleet over the balance of this decade would represent a very promising prospect for substituting domestic natural gas for more-expensive and more-polluting diesel made from imported oil. If U.S. policy begins to stimulate American interest in embracing this approach, the likely environmental and associated economic benefits would be well worthwhile.

It is interesting to see that drilling companies are beginning to convert their rigs to use natural gas instead of diesel fuel. This will not represent a large new market for natural gas. However, it does confirm that the oil industry is itself embracing the economic and environmental advantages afforded by this cheaper and cleaner fuel. As reported in the *Wall Street Journal*, the three largest providers of oilfield services (Schlumberger, Halliburton, and Baker Hughes) are making significant investments to retrofit pumps and rig engines to run on natural gas. This trend appears to be in response to the desires of service company customers ranging from Apache Corporation and Chesapeake Energy to Exxon and RD Shell. They are intending to capitalize on recent advances by manufacturers such as Caterpillar in building heavy equipment that operates on natural gas. From the standpoint of upstream operators, the use of natural gas–fueled drilling and other oil-field service equipment is the noteworthy reduction in emissions versus those associated with the same machinery fueled by diesel. This benefit can be an added plus in obtaining local support for necessary work permits.

In another sign of the broadened use of natural gas, Noble Energy has announced plans to build a plant in Colorado to produce liquefied natural gas (LNG). The company indicated that it plans to use the LNG from this plant to fuel the company's drilling rigs and other equipment in the Denver-Julesburg basin and to provide it as a fuel to other energy companies in the area. Another innovation is at an even earlier stage, as the *Wall Street Journal* has reported that General Electric and Caterpillar are developing new locomotives for trains to also be powered by natural gas. In cooperation with these companies, BNSF (now wholly owned by Berkshire Hathaway) has indicated that it plans to test using LNG as a fuel. The advantages of such a move could include better compliance with tighter EPA emissions regulations that take effect in 2015 and possibly lower costs for an enterprise whose daily fuel bill is $31 million. Thus, BNSF's annual fuel bill amounts to over $11 billion. Clearly the incentive to lower fuel costs while also achieving significant reductions in emissions could be compelling.

The benefits of domestic job creation, reduced CO_2 emissions, and an improved trade balance appear to be carrying the day with respect to U.S. policy on natural gas development. There remain some open issues that the environmental opposition is making, and that will need to be addressed on a continuing basis. The Sierra Club and many other such opponents contend that the greenhouse gas emission benefits from gas development are overstated, because of the problem of fugitive emissions during production and distribution of natural gas. Because methane is considered to be manyfold more powerful as a GHG than is carbon dioxide, the argument is made that such leaks essentially offset the benefits of gas substitution for coal in electricity generation. However, this conclusion is very sensitive to what one believes (or asserts) the percentage of methane leaks to be. The petroleum industry has vigorously asserted that a study by Cornell on this subject is seriously flawed, in that it significantly overstates the magnitude of leaks. MIT in its 2010 study of natural gas acknowledged, "Methane emissions in natural gas production, transportation, and use are not well understood." The MIT study group recommended that the DOE and EPA co-lead an effort to investigate this issue more completely. Because fugitive methane emissions are the loss of a valuable (and otherwise saleable) resource, solutions to the problem are likely to have favorable economic consequences. In any case, addressing what can be done and is being done to minimize such losses to the atmosphere will be critical.

Finally, there have been relatively few, but still concerning, instances of wastewater disposal triggering fairly small earthquakes in Ohio and several other states. To date, this has not been a widespread phenomenon, but it will require monitorship and assessment to determine mitigation options.

The Onset of a New Era for Oil

Mobilizing "tight" oil.

As the petroleum industry experienced a series of successes in unlocking new sources of methane molecules (called "dry" natural gas) from shale source rocks, with only a modest lag, other parallel and very positive developments have begun to occur involving new "unconventional" sources of oil: condensate and NGLs. Foremost among these has been the development of new oil production from the Bakken Formation in the Williston basin of North Dakota and Montana. This new output, and corresponding reserve additions, has grown from a minor level in the late 1990s to an impressive 700,000-plus barrels per day. While this so far represents only a few percentage points of U.S. consumption, it is enough to have recently made North Dakota's 900,000-plus barrels per day of total output the second-largest state in oil production, with a distinct prospect of even higher output over the balance of this decade. Thus, the production outlook is very encouraging for a now substantially validated area comprising almost 15,000 square miles in the Bakken–Three Forks region. At year-end of 2012, already more

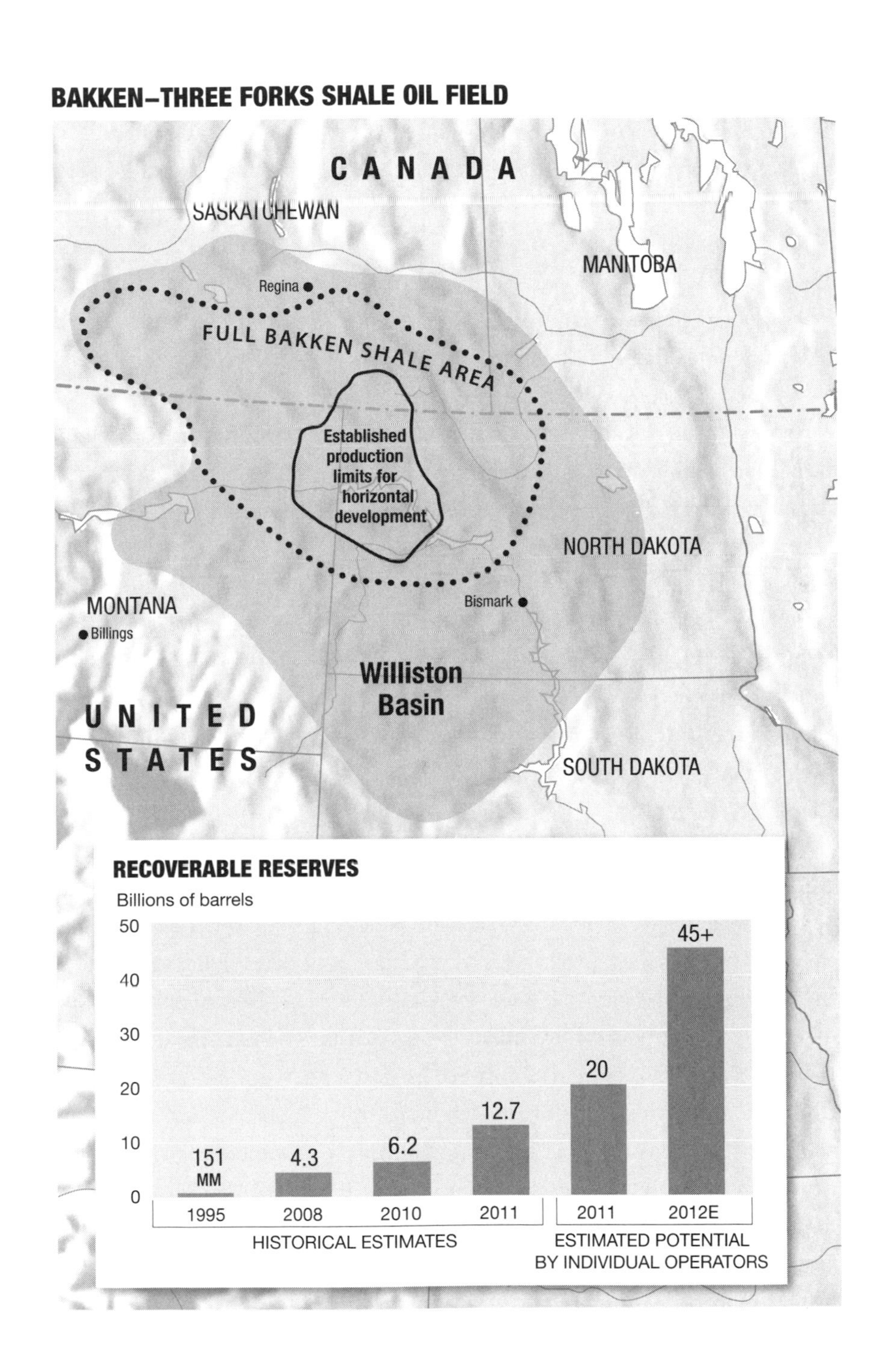
BAKKEN–THREE FORKS SHALE OIL FIELD
CANADA
SASKATCHEWAN
MANITOBA
Regina
FULL BAKKEN SHALE AREA
Established production limits for horizontal development
NORTH DAKOTA
MONTANA
Billings
Bismark
Williston Basin
UNITED STATES
SOUTH DAKOTA
RECOVERABLE RESERVES
Billions of barrels
50
40
30
20
10
0
151 MM
4.3
6.2
12.7
20
45+
1995
2008
2010
2011
2011
2012E
HISTORICAL ESTIMATES
ESTIMATED POTENTIAL BY INDIVIDUAL OPERATORS

than five thousand horizontal wells had been drilled in the basin. Predictions that Bakken–Three Forks production could range between 1.5 million and 1.7 million barrels per day during the second half of this decade seem realistic.

Much of this optimistic view was first articulated at the North America Prospect Expo (NAPE) Conference held in Houston in February 2011. Jack Stark, senior vice president for exploration at Continental Resources, outlined this assessment by the play's leading proponent. For further perspective, he pointed out that in Continental Resources' view, the ultimate oil recovery for the field could be on the order of 20 billion barrels, with additional associated natural gas reserves equivalent in heating value to another 4 billion barrels of oil. An accumulation of this magnitude is truly world scale in its significance. At this level, it would represent almost twice the expected ultimate recovery of Alaska's giant Prudhoe Bay field. In fact, Stark also observed that it is probably the most important new accumulation of oil to be discovered and brought onstream anywhere in the world in more than thirty-five years. A few other larger fields currently have development projects as large as or even larger than the Bakken, but these are fields that were discovered in OPEC countries prior to 1975 and were not scheduled for economic development until just recently with the advent of higher oil prices.

Continental Resources was founded by Harold Hamm, who was born the thirteenth child of a family of sharecroppers in Lexington, Oklahoma. He started in the oil business in the late 1960s, taking small deals and drilling modest wells as was appropriate, given oil prices of three dollars per barrel or sometimes less. Following some exploratory success, Hamm gravitated to an optimistic view somewhat similar to that of Mike Halbouty: both believed that the caution and even pessimism of M. K. Hubbert and his protégés was overdone. Accordingly, at a time when many in the industry decided to pursue easier-to-find natural gas prospects, Hamm's team elected to stay focused on finding and developing oil projects. The Bakken–Three Forks region became their big opportunity. Following an early conventional exploratory venture along the Cedar Creek Anticline near the Montana/North Dakota border, Continental became involved in developing the Bakken Formation in a field called Elm Coulee. The Continental Resources exploration and production team then began to recognize that this success had broader implications for Bakken development over much of the basin. Accordingly, the company began an aggressive lease acquisition program, thereby becoming the largest leaseholder in the basin, with some 1.2 million acres.

At the August 2012 EnerCom conference held annually in Denver, Colorado, for energy investors, Continental Resources' president, Rick Bott, presented an update to Jack Stark's assessment of the Bakken–Three Forks potential. He indicated that deeper drilling through the underlying Three Forks Formation had identified three additional benches or potentially productive layers of rock, two of which yielded core rock samples closely resembling the already productive first horizon of the Three Forks. He cautioned that these additional zones had not yet been flow tested but that Continental was optimistic that they could perform similarly to the first bench of the Three Forks formation. In subsequent announcements, Continental indicated that two of the newest zones had been successfully flow tested and the last zone will be tested shortly. The company's updated assessment reflects that a total oil-in-place potential in the thermally mature region of the Bakken–Three Forks Formations could be as much as 900 billion barrels. Interestingly, this level is nearly twice the number for the yet to be discovered in-place hydrocarbon resource potential cited by Mike Halbouty in 1980 (as described in chapter 3). Bott went on to observe that a recovery of 5 percent from such an in-place resource could ultimately represent a reserve potential of 45 billion barrels. In an updated May 2013 assessment of the Bakken–Three Forks potential, Jack Stark reiterated the company's confidence about a 24 billion barrel oil equivalent reserve potential but also alluded to the even greater upside possibilities depending on recovery rates and ultimate well densities in the core area of the field. This view was reiterated by Rick Bott at EnerCom's Denver Energy Conference in August 2013.

As impressive as the foregoing statements are, there are a few additional qualifiers required to describe the prospectivity of this gigantic Williston basin development opportunity. First, even though the potential reserves are very large, the individual Bakken wells differ significantly from those that are typical of other conventional world-class reservoirs in the Middle East and elsewhere. Current expectations for individual Bakken wells typically are for the recovery of between 400,000 and 700,000 barrels, depending on the thickness of the formation and other factors in a given area. In all, far more than 50,000 wells will probably be required to achieve the ultimate potential reserve recoveries. The capital required just to drill these wells could well exceed $400 billion. Related infrastructure needs could almost double that amount. This implies that the Bakken–Three Forks development will be very capital-intensive and will require multiple decades for full project development. The industry's performance in 2009 suggests that if oil prices were to decline to about $55–60 per barrel on a sustained basis, much of the Bakken development would be only marginally economic. That scenario does not seem likely, but the economic sensitivity to oil prices is nevertheless noteworthy. As another point of caution, some other operators believe that more drilling and testing of the Three Forks benches over a wider area is needed to confirm the consistency of rock attributes. Thus, the ultimate pace of Bakken–Three Forks development will depend in part on future oil prices as well as yet to be completed expanded drilling.

A second caveat involves the infrastructure build-out required to transport Bakken oil to markets that can use it. This has already been a challenge, as the growth in Williston basin production over the past half decade has consistently surprised on the upside. New oil-handling capacity involving both treating facilities and field-gathering pipelines connecting to long-haul pipelines will need to be installed. One example is the announcement of a Williston basin joint venture in oil field-gathering operations by First Reserve Corporation, a private equity enterprise specializing in energy infrastructure investments. Other investment groups are quickly following. When the expected increase in Williston production is combined with projected increases in Canadian oil sands exports to the United States, it is clear that very large additional oil shipping capacity will be required. This suggests that what may be needed is the capability of both the Keystone XL Pipeline as well as the BNSF Railway and other railroads, each of which involve plans for facilities, to move one million or more barrels of oil per day.

In light of this outlook, we should review the process by which this fortuitous set of circumstances has unfolded. In some respects, the raw geologic merit of the Bakken Formation is not an entirely new insight. The Williston basin was opened as a conventional oil-producing region in 1951 by Amerada Petroleum. Early on, the Bakken Formation was recognized as the leading source rock for the Williston basin in which oil was generated and then migrated into a variety of more-porous and permeable rocks from which conventionally trapped economic oil could be produced at prices as low as three dollars per barrel. A more complete understanding of this resource potential largely developed as a result of drilling over a period of thirty or so years along the major north-south-oriented structural feature known as the Nesson Anticline. Subsequently, other seismically defined structures elsewhere in the basin were tested, some successfully and some not. Over time as the presumed "dry holes" (or uneconomic wells) were drilled, the industry developed a technique for completing many of these marginal wells in what was termed the Bakken "bail-out zone." The bail-out idea was based on the notion that although such wells would not recover enough oil to pay out the full cost of drilling the vertical well, if they would produce 10,000–40,000 barrels, such levels would be enough to reduce somewhat the final net cost of the "unsuccessful" projects.

In 1972, the North Dakota Geologic Survey published a book entitled *Near Hits ??? in North Dakota,* by Mary Woods Scott. It contained electrical well logs of so-called dry holes where shows of oil had been detected in the Bakken and other formations. The database of these drilled unsuccessful locations offered important clues as to where future explorationists might unlock economic hydrocarbons. The value of this data became clear as the oil industry began to utilize its newfound ability to drill horizontal wells in relatively tight (that is, low porosity and tight permeability) rock to open up a much larger surface area to contribute to production. By the late 1980s, Burlington Resources and other companies began to pursue new development in one particular zone of the Bakken Formation in Montana. That development was successful, but only moderately so. In retrospect, it is now evident that another part of the Bakken Formation had a much greater potential, but this was unrecognized at the time. Also, the technical ability to drill, stimulate, and complete horizontal wells was still fairly embryonic compared to today. The tasks of keeping the drill bit in the zone of productive hydrocarbons while drilling over much longer horizontal distances and then selectively fracking the rock in multiple stages over distances of one to two miles had yet to be adequately developed. Not surprisingly, early Bakken drilling activity trailed off in the later 1990s, especially when oil prices declined sharply.

However, after the remarkable recovery of oil prices from the 1998 oil price collapse, a new wave of exploration began. It was further stimulated by a series of improvements in horizontal drilling techniques as well as much-enhanced 3-D seismic surveys to better define and then avoid fracking near natural faults. In addition, a more effective well stimulation and completion process evolved (by trial and error) called multistage fracking. At Petrie Parkman, we had a firsthand view of this new development phase when we represented Lyco Energy in the 2004 sale of its newly validated and only partially developed interests in the Elm Coulee field in the eastern Montana portion of the Williston basin. This project evolved from a joint venture research and development project between Lyco and the Halliburton Company. That project's success led many other operating companies to accumulate acreage positions to exploit the Bakken over a much-expanded region of the basin. Among others, these included Hess Corporation, EOG Resources, Marathon Oil, Brigham Exploration (now Statoil), ConocoPhillips, Whiting Petroleum, XTO Energy (now Exxon), and Continental Resources.

Extensive land acquisitions have been followed by active drilling programs by these and other operators. About 180 rigs have been continuously engaged in developing the Bakken–Three Forks opportunities, and there are indications that this could expand by perhaps another 10 percent in 2014. In fact, more rigs are dedicated to this development project than to any other in North America. The Bakken–Three Forks opportunity may well be unique in terms of its size and ultimate production potential for oil. Nevertheless, the success in the Williston basin has set off a series of ventures to unlock other rich hydrocarbon liquids possibilities elsewhere in the United States in a variety of other hydrocarbon basins.

Encouraging progress has already been achieved in the Eagle Ford Formation of South Texas, which was opened by a Petrohawk Energy discovery in 2009. Like the Bakken, this prospective trend covers a very large area. In contrast to the Bakken, it is divided into three distinct producing regions (or windows) with decidedly different economic characteristics: a dry gas window in the deeper portions of the trend, a condensate-rich window in the intermediate depths, and an oil window up on the shallower shelf of the play.

The keys to unlocking the lower energy reservoir of the oil prospective shelf of the Eagle Ford are at an earlier stage of evaluation, and the dry gas

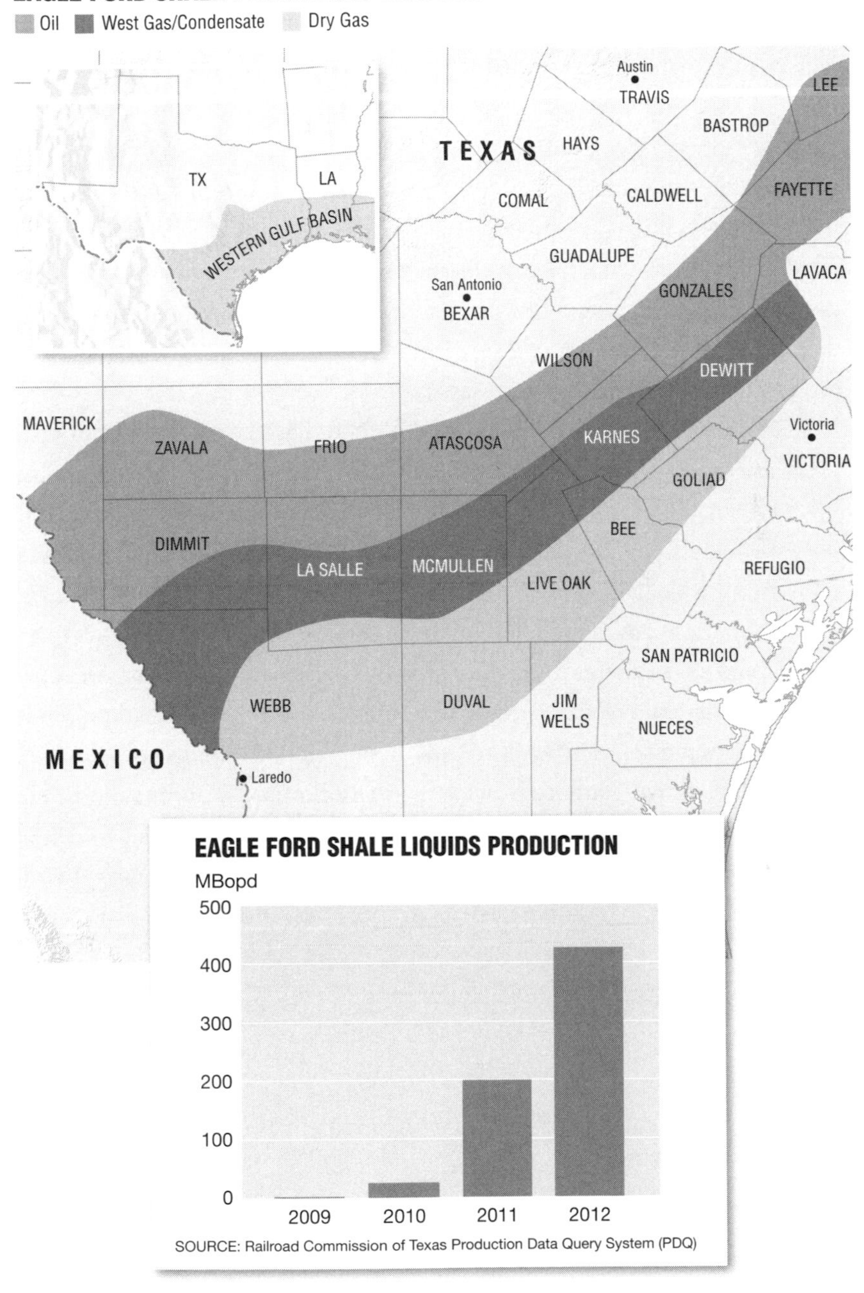

SOURCE: Railroad Commission of Texas Production Data Query System (PDQ)

portion is of less interest as long as natural gas prices remain low. At this point, the most prospective portion of the trend involves the condensate window. Condensate is a very high gravity (that is, light) hydrocarbon liquid that can be thought of as almost naturally occurring motor fuel. While the aggregate reserve potential of the Eagle Ford is most probably not as large as that of the Williston, it is still very encouraging and important. Given the high flow rates of Eagle Ford condensate wells, the payout periods and rates of return on this development generally exceed some of the better Bakken projects. This conclusion has been indicated publicly by Andrew Bryne, director of equity research at IHS, a leading third-party provider of data and economic perspectives on energy issues.

Most of the optimistic cases I describe here have come to fruition only over just the past half decade. Additional drilling over the next five years will be critical to defining more completely the scope and character of the onshore liquids development opportunities in the United States. However, even now it is becoming possible to envision a new era for U.S. oil, condensate, and natural gas liquids development extending beyond the end of this decade and perhaps out to 2030 or possibly even later. Furthermore, over at least the balance of this decade, the potential of the Williston and Eagle Ford development projects affords the first possibility in more than forty years of the United States fully arresting its declining oil production for an extended period and transforming its outlook into a multiyear scenario of sustainable moderate growth. How much further production growth will occur depends on how these horizontal wells perform over their lifetimes. Furthermore, there are other efforts to unlock oil- or liquids-rich potentials in the numerous new horizons already being opened in the Permian basin of West Texas and the Niobrara and Codell Formations of Colorado and Wyoming. There are also possibilities for growing production from the Monterey Shale of California, Ohio's portion of the Utica Formation, and a half-dozen or more other formations involving shales or other source rock in producing basins of Alabama, Mississippi, Louisiana, and other states of the Rocky Mountains, the upper Midwest, and Appalachia.

Another very interesting possibility involves the potential that could be unlocked from source rocks for oil on Alaska's North Slope in the vicinity of the giant Prudhoe Bay field. Some observers have noted the potential for such a development to extend the economic life of the Trans-Alaska Pipeline System. While this idea will require considerable additional capital and field-testing to be fully validated, the unusually large size of the targeted oil in place in the source rock may well justify the effort. In addition, Alaska is exhibiting a more welcoming state regulatory climate.

Much interest is now developing in other parts of the world to pursue the kinds of geologic and engineering ideas that have been pioneered in the United States to unlock unconventional oil. There are now under way a half-dozen or more high-profile joint ventures in U.S. unconventional oil and NGL-rich projects involving national oil companies including those of China, India, and South Korea, as well as other international oil companies such as Statoil, Total, and ENI. These entities are bringing financing for the projects and in return hope to acquire special knowledge and experience that can be applied to similarly prospective formations elsewhere in the world.

It is critical to recognize the potential importance of shale oil development for the U.S. strategic and economic positioning over this decade and the next one to come. There is a good possibility that shale oil projects will grow to a level of 3–4 million barrels per day by 2020 or shortly thereafter. Already it is evident that the Bakken–Three Forks and Eagle Ford projects will each exceed the million barrel per day level of production before mid-decade, with solid prospects for additional growth beyond that level. There are prospects of another million-plus barrels per day of growth in syncrude supplies from the Canadian oil sands and the improving possibility for resumed production growth from deepwater oil development in the Gulf of Mexico. Thus, the U.S. Midcontinent region, from the western slopes of Appalachia to the eastern foothills of the Rocky Mountains to the Gulf of Mexico, has the potential to be among the most liquid-fuel-secure regions in the world for the next several decades. This would provide the United States with assured oil availability for enhanced economic growth out to 2035, a timeline competitive with any region globally.

The *World Energy Outlook 2012* by the IEA confirmed a view similar to this optimistic outlook for the oil sector, with the assertion that U.S. oil and liquids production could rise to 10 million barrels per day in 2015 and advance further to over 11 million barrels per day in 2020. The media reports quickly noted that the total petroleum liquids production level in 2020 could even modestly exceed the expected output levels for Russia and

U.S. OIL PRODUCTION BY TYPE IN THE NEW POLICIES SCENARIO

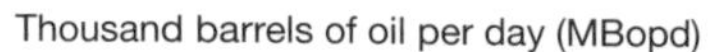

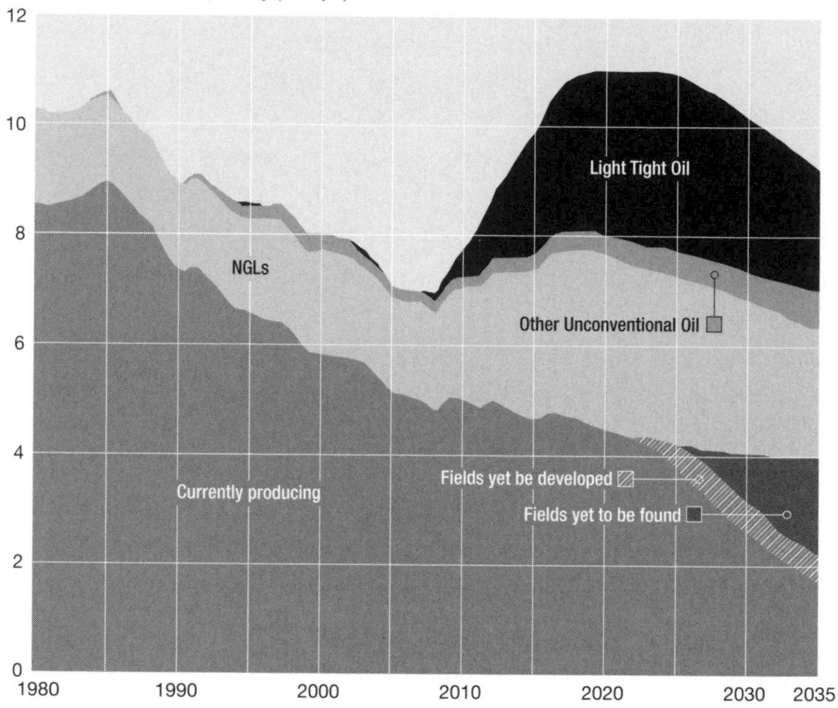

SOURCE: IEA, *World Energy Outlook 2012.*

Saudi Arabia. While this is true, the total for U.S. petroleum liquids for 2020 includes over 2 million barrels per day of lower-value NGLs, which is probably proportionally more than is the case for liquids production by either Russia or Saudi Arabia. In sum, U.S. oil import dependence is likely to diminish substantially over the balance of this decade and could remain at reduced levels well into the next one.

Chapter 9

UPCOMING ISSUES AND CHALLENGES IN THE U.S. ENERGY OUTLOOK

Winning with Effective Leadership

Hope is not a plan.

In June 2006, Matthew R. Simmons (now deceased), an energy investment banker and the founder and chairman of Simmons & Company International, gave a presentation to the Department of Defense in Alexandria, Virginia. This talk was part of the "Energy Conversation Series," designed to focus the U.S. Defense Department and Pentagon strategists on the key drivers of the global oil outlook and other energy supply issues. The title of Simmons' paper was "The Energy Crisis Has Arrived." This and many of his other presentations around the middle of the previous decade were viewed by some as either too extreme or at least cases of seeing the future too soon. Notwithstanding his tendency toward periodic excess, Simmons was determined and articulate in identifying and deepening both the petroleum industry's and the general public's awareness of the formidable challenges involving both maturing U.S. and global production and the scale of growth in energy consumption that is expected to occur across the global economy.

Beyond the clear need to address the rate of its debt issuance, the top challenges facing the United States over the balance of this decade involve (1) continuing to improve our security of energy supplies; (2) making health care more affordable; (3) broadening access to high-quality education; (4) instituting effective immigration reform; and (5) achieving an adaptive twenty-first-century strategic military capability. Furthermore, there is an inherent "taxation equivalent effect" of higher oil and other energy prices on consumer behavior. Thus, a failure to craft an optimal national strategy for domestic energy development could well preclude, or at least restrict, America's flexibility and limit the availability of financial resources needed to address the other four now widely acknowledged priorities. Conversely, execution of an effective program for growing domestic energy production would provide major benefits in terms of increasing tax revenues and adding hundreds of thousands of jobs for petroleum workers in the private sector, with a powerful multiplier effect across other sectors of the economy. A recent example of the latter is the previously cited significant enhancement of the competitiveness of the U.S. petrochemical industry. Furthermore, the emerging signs of a rejuvenation in American manufacturing capabilities are due in no small part to improved security and costs of energy supplies (especially natural gas).

In April 2011, Dr. Joseph Mason, who teaches banking at Louisiana State University and is a senior fellow at the University of Pennsylvania's Wharton School of Business, wrote a *Wall Street Journal* op-ed column criticizing the Obama administration's energy decisions up to that point. Mason's focus was on the policies and actions that were at the time driving deepwater rigs away from the Gulf of Mexico, with adverse effects on job creation and the trade balance. A broader examination makes clear that the federal government's overall range of policies, procedures, and other regulatory decisions in recent years has swung decidedly against domestic development of almost all fossil fuels. With cabinet-level changes, there appears to be the prospect of a somewhat different approach as President Obama's second term continues to unfold. In any case, the fact remains that we still need reliable oil and natural gas supplies to effectively navigate our transition to a future involving a different mix of energy supplies. Let us develop it wisely to get to where we need to be both economically and environmentally.

The cost of imported oil in recent years reached a point at which it has been hampering the U.S. global financial position. Past trends of growing oil import dependence have periodically even raised questions about the credibility of the dollar as a reserve currency. For example, at an import level of 8 million barrels per day and an oil price of $100 per barrel, oil imports account for almost $300 billion of the U.S. trade deficit. Thus, in less than a half decade the cumulative total is well into the trillion-dollar-plus range. In sum, the prospective effects of higher-than-necessary oil imports on the U.S.

trade deficit are concerning. It follows that high dependence on foreign borrowing can adversely affect America's global strategic position.

Fortunately, it is already becoming clear that the United States does have viable options to mitigate and even reverse these trends. Fully accomplishing this transformation will necessitate substantive changes in private-sector as well as governmental leadership. For example, sustained, effective programs to educate the public on energy issues will continue to be required. Such efforts must address the need for responsible domestic energy development and a careful and informed decision-making process to weigh the full extent of the consequences of energy development (or nondevelopment) decisions. Former Shell USA president John Hofmeister's book, *Why We Hate the Oil Companies*, represents a sound start toward the goal of more widely informing the public on U.S. energy issues. He has established a nonprofit foundation, Citizens for Affordable Energy, to advance this effort even further.

Crafting a Pragmatic, Focused Approach to Future Energy Policy

Leadership in energy is as much about actions one elects not to pursue
(because they may be ill-advised)
as it is about what one actually does seek to accomplish.

A pragmatic, focused approach to future energy policy can make a substantive positive difference and, alternatively, avoid many of the adverse effects on U.S. energy security that have been predicted for the coming decade. Specific examples are cited to identify where our approaches to addressing the challenges have been suboptimal.

Let Markets Work, in Their Own Time and at Their Own Pace

The U.S. production outlook for oil and petroleum liquids is now experiencing the best prospect for expansion in more than thirty years. Not since the late 1970s startup of Alaska's Prudhoe Bay field has there been a predictable multiyear outlook for domestic production growth. Furthermore, given the diverse nature of the unconventional oil- and liquids-prone shale resource opportunities, there is the distinct possibility for an extended period of new project developments involving multiple geologic trends in a wide variety of domestic basins. To put this into perspective, U.S. oil production has been in virtual continuous decline since 1970, with the exception of the impact from growing Alaskan production for several years in the late 1970s and very early 1980s. Thus, a peak U.S. oil output of almost 10 million barrels per day had shrunk to a recent low of under 6 million barrels per day. However, overall U.S. production is now experiencing a sustained upward move. This will involve Bakken oil production from the Williston basin virtually doubling from recent levels to a peak of perhaps 1.5 million to 1.7 million barrels per day. In addition, impressive production growth is in store for the Eagle Ford (Texas) and Niobrara (Colorado) Formations along with a strong rejuvenation of production from the giant Permian basin of West Texas. As described previously, there are also numerous basins that can contribute on the margin to this favorable trend.

The case becomes even more compelling if resumed deepwater exploratory activity in the Gulf of Mexico continues to be allowed—and better yet, encouraged. After a couple of very difficult years following the Macondo tragedy, drilling activity in the gulf has recovered substantially. In this regard, the remaining unexplored resource potential of the U.S. deepwater gulf affords an upside approaching the reserves of the now-well-defined U.K. North Sea. This deepwater U.S. opportunity is too important a national treasure to fail to develop with the responsible and safe procedures that are available today. The combined total production gains to be had from each of these sources depend on how a variety of factors affect future development plans. Nevertheless, I am convinced that a new resource potential of sufficient magnitude to raise domestic oil and liquids output back to more than 10 million barrels per day is achievable. Furthermore, an ultimate target of another 10 percent or somewhat higher is possible. Assuming future average oil prices in the range of $80 to $100 per barrel, the benefit to the U.S. trade deficit of such increases in liquids output could well range from $90 billion to $130 billion per year. Further expected improvements in vehicle mileage efficiencies could take this total to well over $200 billion.

The IEA has asserted that the surge in North American oil production "will be as transformative to the market over the next five years as was the rise of Chinese demand over the last 15." As Liam Denning of the *Wall Street Journal* has noted, this is quite a claim. For perspective, he cites the role of growth in Chinese oil demand of 5.6 million barrels per day (accounting for fully 36 percent of the global total growth), which occurred while oil rose from $20 to $100 per barrel. Thus, shrinking U.S. production and growing Chinese demand had the "scissors effect" of driving spare global production capacity to a much too tight 1 percent or less. Now, in contrast, the IEA anticipates the spare capacity could grow to a healthy and supply-flexibility-enhancing almost 8 percent by mid-decade. While there remain some challenges and obstacles to realizing this transformation, the case for constructive optimism is sound.

Interestingly, the Saudi Oil Minister Ali Naimi has essentially agreed with this prospect. He announced in May 2013 that the Kingdom was indefinitely deferring previous plans to increase its production capacity by 2020 to 15 million barrels per day from the current level of 12.5 million barrels per day. He further stated that the decision took into account growing production prospects in Iraq, the Caspian Sea region, Brazil, Africa, and North America. While Naimi continued to assert the Saudi capability to do such an expansion when it might eventually be needed, he made it clear that the Kingdom openly welcomes that not to be the case for the foreseeable future. In September 2013, the *Wall Street Journal* reported that Saudi Oil Minister Ali Naimi essentially reiterated this view with an assertion that U.S. production growth has "increased the depth and stability for oil markets."

If the United States proceeds to capitalize effectively on these game-changing developments, the IEA data suggest the clear potential for a large improvement in the U.S. strategic positioning by the latter part of this decade. On that basis, we opened the current decade with domestic oil production of 7 million barrels per day versus consumption of 17.5 million barrels per day. The corresponding oil imports were approximately 10.5 million barrels per day. In recent years, our oil import cost has accounted for between a third and 45 percent of our structural trade deficit. There are many factors in addition to growing domestic oil supply that will determine our future position, importantly including various gains in consumption efficiencies. These savings could range from 600,000 barrels per day in 2018 to over 2 million barrels per day in 2025. Thus, it is not unreasonable to expect that we could close the decade with oil imports of well under one-half of recent levels. This swing would generate commensurate reductions in foreign borrowing needed to balance our trade accounts. Therefore, something on the order of a 30–40 percent reduction in our structural trade deficit could be achievable. The direct and indirect domestic job creation gains from accomplishing this transformation would be substantial. In fact, with a total import bill reduction in excess of $200 billion, these benefits would likely be among the largest to be had by any of the components of the U.S. industrial and manufacturing sectors. In fact, with growing Canadian output, this could be considerably less from non-North American sources.

U.S. dependence on imported oil could well be even further reduced as there is increased utilization of natural gas, especially in the transportation sector. Over the longer term, this could involve both an expansion of natural gas vehicles (in the various modes described in chapter 8) and the possible introduction of electric and hybrid electric vehicles fueled increasingly by natural gas generation of electricity. Furthermore, the financial benefits to the net petroleum trade balance should be improved as the United States builds out its infrastructure for exports of LNG to various international markets. In addition, Tom Donilon, the former White House national security advisor, observed that the United States' new "energy posture allows us to engage [with the world] from a greater position of strength." Increased use of natural gas has already transformed the United States from the leading emitter of carbon dioxide into the largest absolute mitigator of such emissions by an impressive margin over the European total, and further significant progress is likely. As observed by Ed Crooks and Geoff Dyer in the *Financial Times*, "The shale boom will not allow America to disengage from the Middle East, but its progress toward oil self-sufficiency will bolster Washington's position in international diplomacy."

Avoid Flawed Economic and Policy Incentives That Can Cause Unnecessary Supply Constraints

Security of the U.S. oil supply should also be improved by not inhibiting the expanded development of Canadian oil sands resources, given current investment programs designed to almost double current output from 1.5 million by early in the next decade. The key to achieving this benchmark rests on regulatory oversight and required governmental approvals becoming streamlined and predictable so that longer-term capital commitments can be formulated with confidence. This has not been the case recently. The

ILLUSTRATIVE PATH TO U.S. PETROLEUM FLEXIBILITY

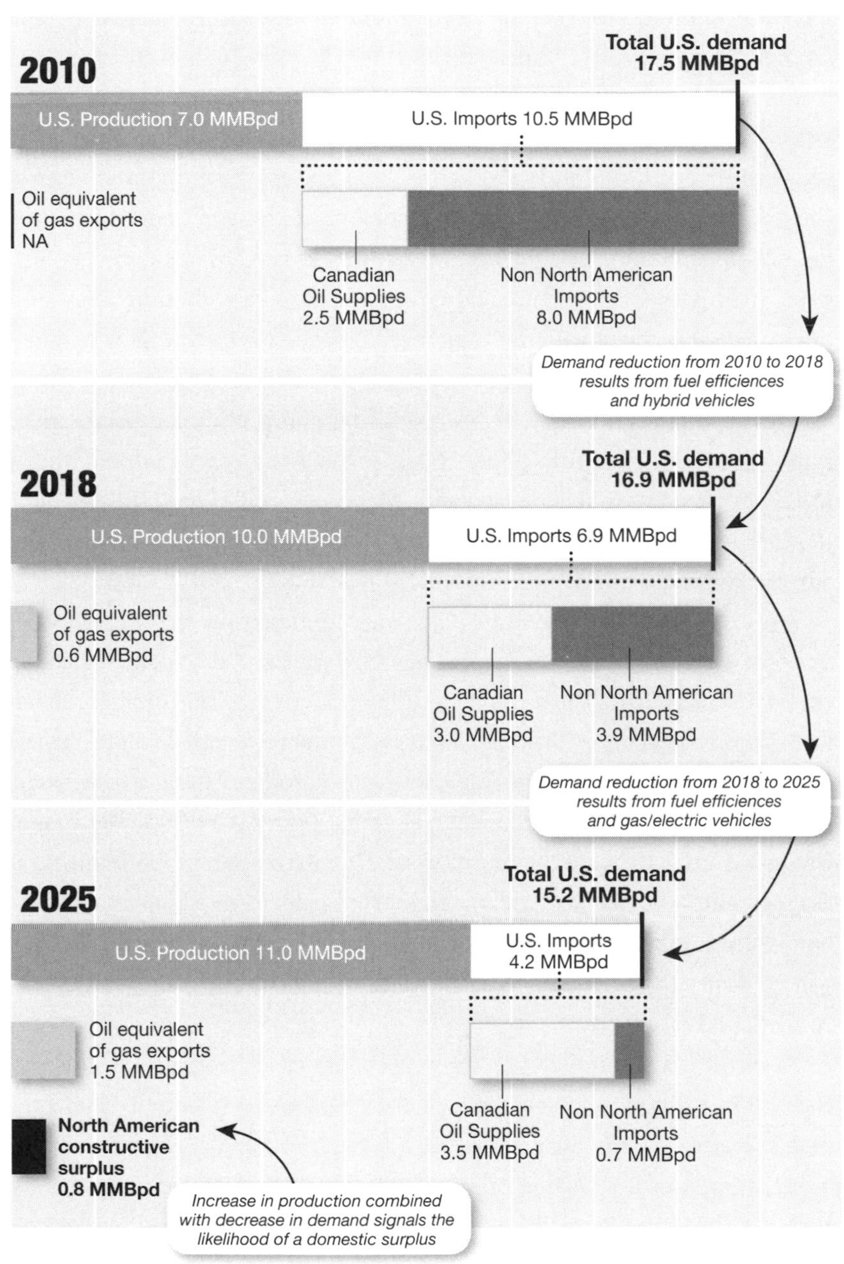

SOURCES: IEA, *World Energy Outlook 2012*, and author's estimates.

Macondo disaster has clearly exacerbated the petroleum regulatory environment and outlook. However, the overreaction to the Macondo disaster is not an isolated example, as outlined below.

Regarding Canadian oil sands, the objections to TransCanada's Keystone XL Pipeline has been particularly multifaceted and strong. As a result, the permit approval process has been stretched out, resulting in additional costs and delays for expected commencement of construction. At best, the final completion date of 2013 has now slipped to late 2015 or quite possibly 2016. In this case, environmental advocates have raised numerous questions, including many with doubtful merit, about the pipeline. However, the openly acknowledged agenda of these groups actually is to undermine future Canadian oil sands development by restricting transportation capacity to the most relevant market, the United States. While their goal is to deny access to U.S. markets, the likelihood remains that the oil will then ultimately flow to other markets in Asia and possibly Europe—ironically, with even greater adverse global environmental impact.

The debate over the Keystone XL project underscores the broader confrontation involving environmental interests opposed to North American energy development. There is evidence from a credible third-party consultant (Cambridge Energy Research Associates) that the environmental impact of oil sands is essentially comparable to other oil sources already being used, including California heavy crude oil, Nigerian light oil, and Middle East as well as Venezuelan heavy oil. The heart of the environmentalists' argument is that extraction of Canadian oil sands produces more greenhouse gases than does conventional oil. What the opposition ignores is that such "syncrude" will be displacing generally comparable heavy oils as opposed to light conventional oil.

Nevertheless, several groups have remained rigid in opposing the project. In the early fall of 2011, the U.S. State Department initially indicated that it was ready to advance the permitting process. However, it subsequently announced a further one-year-plus deferral of its decision for what appear to be domestic political considerations. This is another example of the tendency of political leaders to "kick the barrel" down the road because of perceived environmental sensitivities rather than embrace an effective option for materially improving the security of American oil supplies still with appropriate environmental protections (and even enhancements). It is noteworthy that the abrupt unilateral U.S. decision to delay the pipeline permitting process has motivated Canadian governmental and business leaders to upgrade discussions with China and other Asian countries regarding new infrastructure to facilitate shipping energy and other natural resources to Pacific Rim markets. Finally, in March 2013 (almost a year and a half later), the U.S. State Department once again confirmed that the Keystone XL Pipeline does not pose any greater threat than the alternatives. At this point, the Williston basin's need for this pipeline has diminished because of much-expanded railroad transportation capacity and the appeal of Bakken crude to a wide variety of refiners. Nevertheless, U.S. action that encourages America's largest trading partner to tighten its economic and natural resource relationships with other nations is of dubious merit.

Another reaction by TransCanada Corporation (the operator of Keystone XL Pipeline) that was possibly motivated by the extreme slowness of the U.S. government in approving Keystone XL is a new proposal to build an oil pipeline with a capacity of 850,000 barrels per day from Alberta to East Coast Canadian markets. This would be called the Energy East Pipeline. This may well reflect Canadian desires to provide wider access to alternative markets for their crude oil. As indicated in Canada's federal budget, the oil price declines because of delivery bottlenecks associated with the holdup on permitting the Keystone XL Pipeline and other delayed pipeline approvals may shave as much as 28 billion Canadian dollars off the GDP in 2014. It would reduce actual government revenues by 4 billion Canadian dollars. Furthermore, it represents additional evidence that Alberta oil (including oil sands) is likely to find its way into various markets one way or another.

Another area in which flawed incentives also have been counterproductive involves ethanol mandates. Using ethanol as an oxygenate booster to gasoline is effective and beneficial up to about 10 percent of the gasoline blend. However, state and federal directives to achieve even higher percentages are of debatable value (and actually of doubtful merit) in terms of both fuel performance and unintended adverse consequences for prices of both gasoline and food utilizing corn. Conversely, the recent expiration of ethanol blending credits is a step in the right direction.

Expect and Accept That It Is Often Darkest before the Dawn

As one considers the cyclicality of oil prices depicted in the frontispiece, one can see that self-correcting forces inevitably (and often relatively quickly) become operative when volatility reigns. The best cure for high oil prices is high oil prices, and the best cure for low oil prices is low oil prices. Letting such cycles unfold along their natural course is important and preferable because the value of having price signals to the marketplace is critical in triggering improvements in consumption efficiencies with new innovations in technology, as well as stimulating corrective forces that lead to new supply availability. The most recent examples of this process involve the development of unconventional natural gas and now the shale oil revolution, both of which are proving to be the highly rewarding fruits of a patient policy approach to deal with fluctuations to the high side of oil and gas prices over the past decade. The concerns so evident in 2004 and 2005 regarding energy supply shortfalls emanating from peak oil have probably been indefinitely postponed. This is a lesson learned involving both the need to have faith in the likelihood of ongoing technical advances and the exercise of restraint in reacting to short-term price swings. It is one that will be important to recall and apply repeatedly as future energy challenges inevitably arise.

Recognize That Black Swan Events Often Entail Especially Noteworthy Risks

In the public arena, the exercise of sound judgment in dealing with low-probability but high-impact events is critical. Regarding deepwater oil development in the Gulf of Mexico, after a very confrontational beginning, the Obama administration appears to have demonstrated a somewhat more open and constructive view toward both leasing and permitting prospective acreage. After the Macondo blowout and spill, the administration was justifiably criticized for overreacting by declaring that all new drilling permits would require a new environmental impact statement. The administration has subsequently shown signs of softening its stance, but the prior policy decision may still have facilitated a basis for even more extended litigated opposition to offshore drilling as well as new offshore leasing.

The Ruby Pipeline permitting experience is another example of excessive regulatory burdens stemming from an objective of avoiding the consequences of black swan events. This is a grassroots project designed to deliver natural gas from Opal, Wyoming, to West Coast pipeline interconnects in Malin, Oregon. The project is a key step in the reconfiguration of the U.S. pipeline grid and is intended to help rebalance supply with demand nationally by facilitating greater shipments of Rocky Mountain gas to West Coast markets. It is both critical and desirable because the likely growth in eastern and southern U.S. shale gas supplies will be ample to meet eastern and midwestern market requirements for multiple decades. In contrast, the level of West Coast supplies to meet its needs is less certain. The project was completed about four months behind schedule and 23 percent over budget. This outcome reflected more than 125 stakeholder meetings and various agency scoping sessions. In addition, there were requirements involving a whole range of considerations for migratory birds, other wildlife, and thousands of Native American artifacts and other cultural resources. Numerous individual lawsuits have been filed one after another, not all of which are yet fully resolved. The Ruby Pipeline experience is instructive in that building and reconfiguring a series of similar significant gas pipeline infrastructure projects will be essential over the next two decades to optimize exploitation of the shale gas opportunities.

Beyond the matter of regulatory burdens in constructing new pipelines and related infrastructure for new development, there is ongoing opposition by environmental groups to the use of fracking to develop oil and gas in shales and other tight, or low-permeability, rocks. In large part, this has been fostered by often misleading and sometimes exaggerated assertions about the risks of such projects contaminating aquifers that supply water for human use. For example, the movie *Gasland* heightened such concerns by showing a water faucet that burned like a blowtorch when turned on and lit. A similar scene is contained in the movie *Promised Land.* According to Lachlan Markay of the Heritage Foundation, that movie was significantly financed by a media company owned by the government of Abu Dhabi. The latter country along with other members of OPEC and Russia have expressed their opposition to the unconventional resource revolution, apparently because it is considered a threat to demand for their exports. However dramatic the "flaming faucet" incident depicted, the appropriate regulatory authority has acknowledged that it actually involves a water source long known to contain natural gas and not one related to pollution by fracking in the first case. In the other case, a court is reported to have found that there was fabricated injection of natural gas into a waterline to simulate the alleged situation. In contrast to these movies, a balanced and comprehensive discussion of the fracking process and related issues can be found in the book *Fracking: America's Alternative Energy Revolution* by John Graves.

To be sure, there are water wells in Appalachia and Colorado and many other hydrocarbon productive regions in the United States that contain samples of methane. This phenomenon can occur as a natural migration of hydrocarbons into an aquifer via either faults in the earth's crust or, in a few cases, a casing pipe failure in an old producing well. For this reason, it is becoming standard industry practice in these areas to do a baseline survey of water samples around newly proposed well sites to determine any preexisting conditions of methane presence. In addition, because the failure of older casing head pipe has been known to occur occasionally, the design of new wells with even higher integrity standards is well recognized as a critical preventive measure by both operators and state regulatory agencies. Because of this pattern of apparently misleading assertions, it would not be surprising to hear of other fear-inducing claims associated with fracking. Post-Macondo, the petroleum sector is appropriately sensitized to its responsibilities to strenuously avoid incurring adverse environmental impacts. Full elimination of such risks may not be possible, but effective contingency planning to minimize the consequences is a critical and thus high-priority objective of responsible operators. The tragic July 8, 2013, derailing of oil tanker cars at Lac-Megantic in Quebec further underscores this conclusion.

Embrace the Powerful Regenerative Forces of Incentivized Capital

Each of the fracking-associated issues described here as well as others were identified and analyzed by the Shale Gas Subcommittee set up by Stephen Chu, then Secretary of Energy. The committee consisted of prominent citizens representing academia, environmental interests, and the petroleum advisory sector. They found that fracking techniques are vital in the development of unconventional resources. The panel noted that fracking needs to be utilized with a high degree of care to achieve best-practice operating standards. Still to be determined by the executive branch is the question of continuing the current practice of state regulation of fracking operations or, alternatively, imposing regulations on a federally administered basis. Not surprisingly, both the industry and many of the states involved strongly prefer the current approach and argue that they are better positioned to make appropriate local determinations reflective of the specific formation characteristics in their territory. The EPA is believed to advocate for federally promulgated standards. If so implemented, this would undoubtedly involve considerable duplication of effort, because full elimination of state oversight is both unlikely and probably counterproductive. In sum, the argument is that a one-size-fits-all approach is likely to prove suboptimal because of multiple layers of regulatory approvals and mandated, but technically unnecessary, requirements in numerous cases.

In May 2013, the Obama administration issued through the Department of the Interior new proposed rules governing hydraulic fracking on public lands. These rules will apply to an estimated 700 million acres of lands administered by the Bureau of Land Management as well as 56 million acres of Indian lands. These rules continue to allow a limited degree of confidentiality protection regarding certain components of fracking fluids used. However, for the bulk of the fluids involved, there is required public disclosure on an industry-operated website (FracFocus.org). It also provides for the use of well integrity tests for a representative well in a given project, as opposed to requiring all wells to be so tested. Furthermore, on the issue of state versus federal regulation of fracking, it proposes that state regulations can be made applicable in cases where federal officials have determined those standards to be as tough as or tougher than those of the applicable federal rules.

Not too surprisingly, a number of environmental groups have expressed dissatisfaction with this proposal. While much depends on how this rule making is actually implemented, it initially appears to be a somewhat constructive Department of Interior effort under the new leadership of Secretary Sally Jewell, a mechanical engineer by training with some personal experience fracking wells for Mobil early in her career. Nevertheless, it is noteworthy that a coalition of governors of western states has petitioned to be exempted from this rule making, viewing the provision as unnecessarily constraining versus the protections of existing regulations.

The point of all the foregoing discussion is not to argue that environmental and other regulatory provisions should be eliminated. That is clearly an impossibility, and many of the regulations and rules are appropriate. However, if the country's high-potential opportunities to reshape its energy future effectively are to be realized, a pragmatic balance needs to be struck between how we address valid environmental and other local concerns versus how we regulate and effectively allow the private sector to embrace and accomplish what is safely possible in this newly transformed energy supply development. The cases I have cited all relate to the petroleum sector. However, there are equally concerning counterparts to these stories involving the coal industry and the nuclear power sector.

Remain Alert to the Disruptive Potential of Old Geopolitical Grudges

Saddam Hussein's use of the long-disputed claim that Kuwait was the nineteenth province of Iraq to justify his invasion of that country is an example of such a grudge that led to a globally disruptive confrontation. Another involved the Iranian resentment over oil market shares that led to the oil price collapse of 1997–98. Around the world there remain numerous similar unsettled issues and festering situations that inevitably will crop up to pose new challenges in the coming years. Some of these will be related to the emerging power triangles that have developed between Moscow, Beijing, Tehran, and New Delhi. Others will involve the traditionally complex Middle East relationships, such as the more-than-simmering Iranian/Saudi confrontation over Shiite versus Sunni leadership issues in Bahrain. Even more concerning are the issues leading to violence and a breakdown in civil society in Syria, with implications for Turkey, Iraq, Saudi Arabia, Lebanon, Iran, Jordan, and Israel. Similarly, we are already seeing second-order effects of the Libyan revolution that deposed Moammar Khadafy. For example, across North Africa there are cases of renewed terrorism by Muslim militants, especially in Algeria and Mali. Other violence is occuring in Southeast Asia, particularly in Malaysia and the Philippines. Finally, China's rising assertiveness regarding ownership of islands and mineral rights in the South and East China Seas is a serious concern. Rewards will accrue to both nations and investors prepared through contingency planning to deal with (but judiciously not overreact to) the knock-on effects of unsettled grudges.

Capitalize on the Benefit of Periodic M&A Consolidation to Stimulate Innovation

The amazing progress that has been made over the past four decades may very well lead to optimism about U.S. prospects for a viable energy outlook, even looking into the 2030–2050 time frame. What provides particular encouragement and hope for a better future is the overwhelming evidence of the power of the learning curve that is manifest in America's processes for dealing with these issues. Each of the three major M&A-induced corporate consolidations (the early to mid 1980s, the late 1990s, and the first half of the 2000s) unleashed and motivated highly talented people ready to apply their intellectual capital in pursuit of resource development projects from newly capitalized platforms. What remains valid is M. King Hubbert's admonition (quoted earlier) that "our ignorance is not so vast as our failure to use what we know." However, it is also true that there are many examples in which we are now doing a much better job of effectively applying what we know to solving our energy-supply challenges than has been the case historically. The experiences of the many decades recounted here have continually impressed me because they capture the power and ingenuity of our private-sector enterprises to learn from both successes and mistakes and to innovate aggressively in pursuit of sound economic opportunity. Many other parts of the world are once again looking to U.S. technology-based leadership in energy development. This has the potential to be a major advantage for the United States as we confront the significant geopolitical, economic, and environmental challenges of the twenty-first century.

Continually Assess How Shifting Global Macroeconomic Drivers Can Periodically Affect Both National and Corporate Energy Strategies

The recent changes in the outlook for U.S. natural gas provide a valuable example of how dramatic such shifts can be. As described in chapter 6, the private sector with strong U.S. government regulatory support from the

Federal Energy Regulatory Commission (FERC) built out a large new infrastructure to accommodate the importation of liquefied natural gas. These facilities were anticipated to be needed as domestic gas production entered a presumed "inevitable" and "irreversible" natural decline. Yet, they built it and so few came! To date, utilization of these facilities has been only a minor percentage of the available capacity because the new supply from the shale gas revolution has overwhelmed what seemed to be a solid strategy less than a decade ago.

Put another way, if one makes a prediction (or undertakes an initiative), it needs to be reassessed periodically (indeed, almost continually) to take account of changing circumstances in today's dynamic world economy. Accordingly, many of the companies with gas importing facilities are now adroitly pivoting to pursue the exact opposite strategy by exporting some of the emerging surplus in U.S. unconventional gas production. Frequently assessing shifting macroeconomic realities regarding China and other Asian growth, along with trends in a variety of other energy sources, will be an important determinant of U.S. successes in dealing with the challenges of the emerging Middle East/Asian power triangles.

The Six-Day War of June 1967 could have been the first oil crisis. However, such was not to be, because in part the United States had ample shut-in oil production that was quickly brought onstream. Within the global oil supply mix, there also were numerous non-Arab countries able and willing to increase their output to help offset the Arab oil embargo. Until now, that was the last occasion of fortuitous oil supply diversity and flexibility. Over the intervening almost one-half century as global demand increased and as U.S. and other non-OPEC sources of production declined, there have been four major conflagrations that have adversely affected the security of the oil supply and have impacted petroleum prices globally. Numerous other lesser events have periodically threatened smaller but still noteworthy disruptions to the availability and cost of oil supplies. Through it all, the efforts of seven previous American presidents to implement an effective energy strategy never gained much traction. Moreover, on numerous occasions, congressional measures led to demonstrably counterproductive initiatives. On a cumulative basis over the years, the United States and other oil-importing nations have incurred significant economic burdens from these events in terms of wider trade imbalances, inflationary impacts, and larger than otherwise necessary foreign debt issuance.

In a paper entitled "Oil Capacity Outlook for the Arabian Gulf Five" presented in the middle of the previous decade by Dr. Sadad al-Husseini, there was a concluding statement: "The oil industry needs a paradigm shift in terms of reconciling the dynamics of finite oil resources, spiraling energy demand, and the complex political currents within which production strategies are determined." Although the supply-expanding benefits of the shale revolution represent only a partial answer to the global shift for which Dr. Sadad al-Husseini was calling, it does represent a major advance in just that direction. When combined with sensible measures to improve efficiency of energy consumption and cooperative open-market policies internationally, it should lead to a much less dire global supply/demand outlook for the next few decades than was generally warned by the late Matt Simmons and others less than a decade ago.

Winston Churchill is reported to have once remarked, "You can always count on Americans to do the right thing—after they've tried everything else." This comment was originally offered in a rather different military and geopolitical context. It is also now somewhat overused in speeches and other presentations. Nevertheless, Churchill's formulation is now particularly apt in describing portions of the energy picture I have recounted regarding the past four decades. After many missteps and false starts (as well as notable successes), the United States, and by extension many other countries, at last has a workable range of options to pursue doing the right thing on energy.

In announcing the 2012 edition of *World Energy Outlook*, the IEA observed, "The global energy map is changing, with potentially far-reaching consequences for energy markets and trade. It is being redrawn by the resurgence in oil and gas production in the United States and could be further reshaped by a retreat from nuclear power in some countries, by continued rapid growth in the use of wind and solar technologies, and by the global spread of unconventional gas production. Perspectives for international oil markets hinge on Iraq's success in revitalizing its oil sector." It added, "These changes will recast expectations about the role of different countries, regions and fuels in the global energy system over the coming decades."

In contrast to many decades of adverse and challenging oil experiences with limited options, the United States can now strengthen its strategic energy balance during the next several decades. Thus, America's ability to cope with whatever energy-supply interruptions may develop is likely to prove more manageable than has been the case in any other time in recent decades. To be sure, the risks of economically disruptive events occurring have not notably diminished. However, the U.S. ability is now well established to capitalize on the revolution in exploiting unconventional hydrocarbon resources and also pursuing deeper, more difficult conventional resources. If exploited wisely, these opportunities will provide a greater degree of flexibility and freedom in addressing the formidable challenges associated with inevitable geopolitical realignments (such as new emerging power triangles) and other, as-yet-unpredictable upheavals. Importantly, it will provide the much-needed bridge to a future in which hoped-for and emerging advances in both renewable and nuclear energy technologies along with expanded conservation measures can eventually become much larger contributors to America's energy security position.

EPILOGUE

In late 2006, as my partners and I were wrapping up our seventeenth year in business with a record level of transaction advisory engagements, Petrie Parkman merged into Merrill Lynch & Company. In terms of key personnel dedicated to Merrill's North American upstream petroleum investment banking practice, this effectively doubled that firm's capacity in this sector. More than two-thirds of our Petrie Parkman professionals joined Merrill initially.

Following the acquisition of Petrie Parkman, I became a Vice Chairman of Merrill Lynch with client coverage responsibilities for a global range of corporate energy relationships. Accordingly, I began to make frequent trips to the United Kingdom, Europe, Canada, and India. In the fall of 2008, Merrill agreed to be acquired by Bank of America, effective January 2009, and I became a Vice Chairman of the combined entity with a similar scope of responsibilities. Our Petrie Parkman investment bankers became integrated into the Merrill structure and subsequently that of the Bank of America Merrill Lynch (BAML) organization. The burgeoning interest in unconventional gas and oil development projects became a major area of focus for these investment bankers. These included public financings for the Bakken industry leader Continental Resources as well as for other companies pursuing unconventional resource development. It also involved a series of joint ventures for many foreign oil and gas enterprises looking to capitalize on the learning curve experiences of U.S. E&P companies focusing on unconventional projects. In April 2012, having completed a five-year contractual commitment to BAML, I then joined a group of former colleagues at an enterprise now known as Petrie Partners, an energy investment banking boutique providing strategic advisory services to the petroleum sector.

Appendixes

Appendix A

U.S. NATURAL GAS TRANSPORTATION SWITCH AND EFFECT ON EMISSIONS

U.S. NATURAL GAS TRANSPORTATION SWITCH

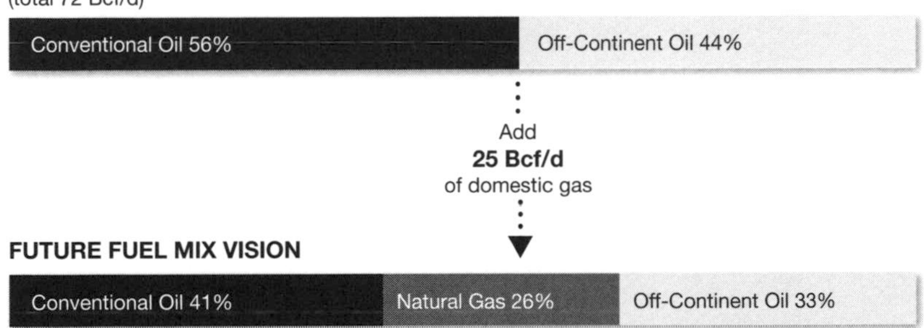

EFFECT ON EMISSIONS

Metric tons

Current Future Reduction

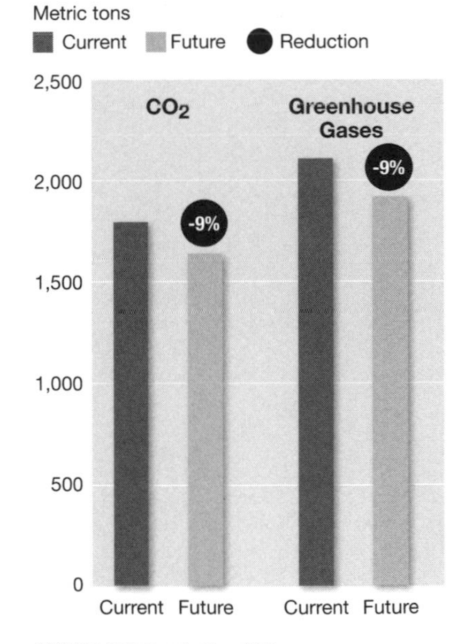

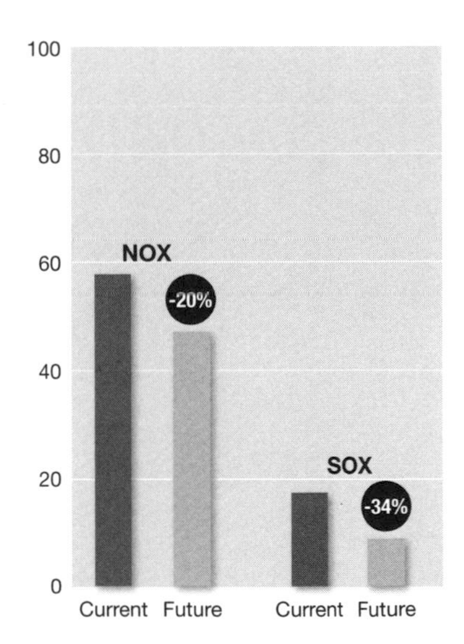

SOURCE: EnCana projections 2008.

Appendix B

WORLD EVENTS AND OIL PRICE TRENDS (1971–2013)

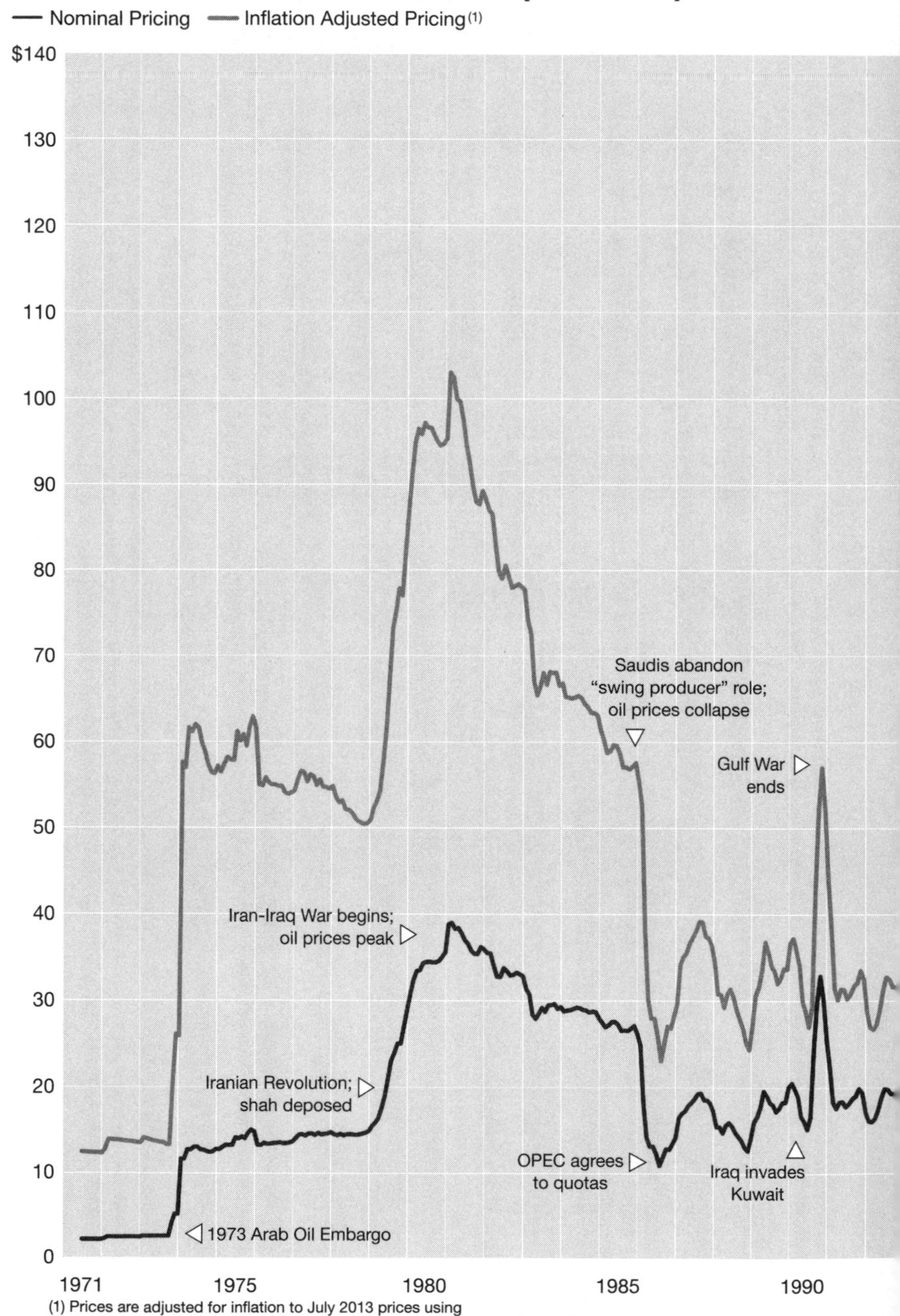

(1) Prices are adjusted for inflation to July 2013 prices using the Consumer Price Index (CPI-U) as presented by the Bureau of Labor Statistics.

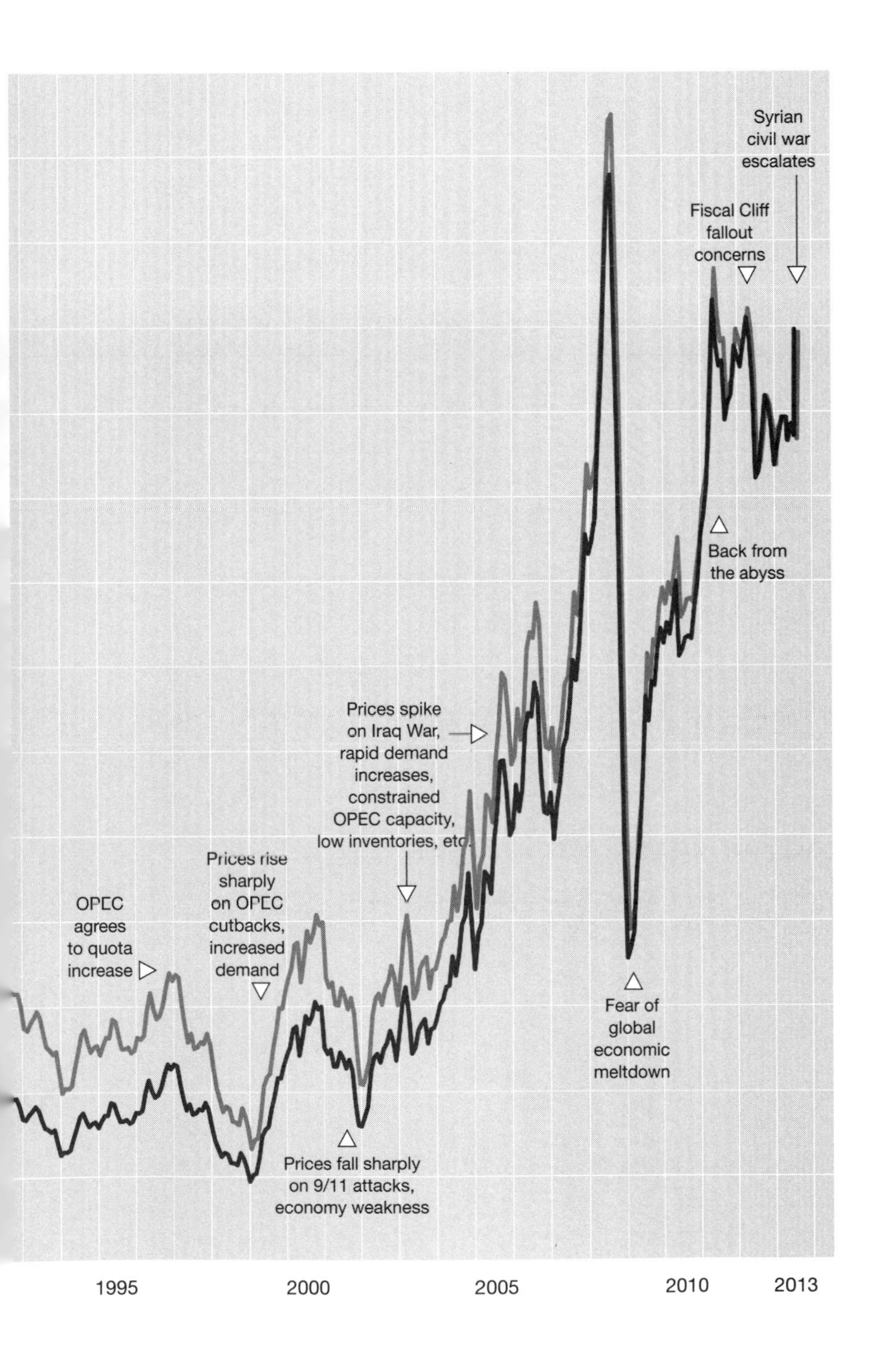
Syrian
civil war
escalates
Fiscal Cliff
fallout
concerns
Back from
the abyss
Prices spike
on Iraq War,
rapid demand
increases,
constrained
OPEC capacity,
low inventories, etc.
Prices rise
sharply
on OPEC
cutbacks,
increased
demand
OPEC
agrees
to quota
increase
Fear of
global
economic
meltdown
Prices fall sharply
on 9/11 attacks,
economy weakness
1995
2000
2005
2010
2013

Appendix C

INVESTING RULES

In 2001, I was asked to participate in the publication of a book by providing a concise set of "rules" that I considered useful in for investing in the energy sector. The book contained a collection of observations by some 150 different investment analysts, portfolio strategists, economists, and other thought leaders experienced in assessing investment options. This was originally published in *The Harriman Book of Investing Rules: Collected Wisdom from the World's Top 150 Investors*. My rendition of these rules (on pages 339–40) addressed issues involving geopolitics, cyclicality, oil and gas prices, financial leverage, quality of management, technological innovation, stock repurchases, and a perspective on alternative energy options. Reproduced below by permission of Harriman House are the rules that I submitted based on my years of experience (thirty years at that time, eighteen of which I was a petroleum analyst and another twelve of which I was an investment banker). Now, over a decade later, I find that they have largely stood the further test of time.

1. **Geopolitics matter.**

   The variable effectiveness and changing policies of the Organization of Petroleum Exporting Countries (OPEC) as well as the major consuming nations can shape the energy sector's overall investment attractiveness.

2. **Always remember that the business is cyclical.**

   While there are longer-term secular trends in terms of demand growth and supply additions, the industry's overall importance to broad measures of economic performance periodically results in pronounced cyclicality.

3. **The best cure for low oil and gas prices is low prices, and vice versa.**

   History shows that $10.00/bbl oil begets $25.00/bbl oil, and conversely, sharp upward moves to $30.00+/bbl oil are usually unsustainable. Low prices tighten supply versus demand; high prices do the opposite.

4. **Contrarians are periodically highly rewarded.**

   When consensus is clearly negative about energy commodities, the stocks are often excellent buys (i.e., 1986 and 1998); when consensus is uniformly positive, watch out (i.e., 1979–1980).

5. **Good exploratory well news travels fast; bad news often seeps out slowly.**

   When assessing the impact of high-potential exploratory drilling, it seems that slowly developing announcements of results seldom match positive expectations. Remember the adage "Buy on expectation potential; sell on actual announcement."

6. **High financial leverage with typically high energy commodity price volatility can be a deadly combination.**

   Most energy company bankruptcies result from ill-timed uses of debt to acquire or develop production in anticipation of a commodity upswing that fails to materialize on schedule. Accordingly, corporate strategies emphasizing financial leverage are often a risky bet.

7. **Quality of management does matter.**

   Significant capital destruction is not an uncommon occurrence among energy companies; accordingly, managements that exhibit consistent financial discipline in the capital allocation process often merit a premium.

8. **Technology counts.**

   Many companies in the energy sector are often viewed as "old economy" stocks. Nevertheless, the role of "new" technologies in unlocking energy resources more efficiently and at lower costs is critical to corporate success.

9. **Stock repurchases deserve attention.**

   In the 1950s and 1960s, J. Paul Getty validated the idea that recapturing the barrels of oil equivalent behind outstanding common shares could be financially preferable to drilling new discoveries. In the 1970s and early 1980s, T. Boone Pickens developed a variation of this idea. Thus, an astutely executed corporate stock repurchase program can provide a useful clue to a company's investment attractiveness.

10. **Beware of popularized alternative energy concepts.**

    While diversification away from conventional hydrocarbon sources is undoubtedly both desirable and inevitable, the path to uncovering profitable and viable alternative energy enterprises is likely to be as tortuous and risky as many other sectors involved in pursuing technological innovation have also demonstrated.

NOTES

Chapter 1. Early Encounters with the Oil Sector

3 Middle East tensions: Associated Press, *Lightning out of Israel,* 43–51; Oren, *Six Days of War,* 61–126.

5 oil prices to more than triple: "Mood of Moderation Prevails at Meeting of OPEC Ministers to Fix New Oil Prices," *Los Angeles Times,* September 25, 1975.

8 the economics of LNG: author's research notes, July 1971.

10 Hess Oil and Chemical merged into Amerada Petroleum: Hess Corporation website, www.hess.com; "Oil Stocks New Interest for the Markets," *New York Times,* March 23, 1969.

12 land grant awards: Ambrose, *Nothing Like It,* 95.

12 as other historians have chronicled: Bain, *Empire Express*; White, *Railroaded.*

12 land itself had limited value: Ambrose, *Nothing Like It,* 376.

15 On March 12, 1968, BP announced: www.bp.com/.../a/A03_prudhoe_bay_fact_sheet.pdf.

18 tanker rates to collapse around year-end. "Fewer Oil Tankers Sail the Seas since the Arab Embargo Started," *Telegraph,* December 28, 1973.

Chapter 2. A Tumultuous Decade

21 inflation having risen: "Rate of Inflation Continues Higher Than Nixon Goal," *New York Times,* November 22, 1972.

21 Congress, the Executive Branch and the courts devoted: Vietor, *Energy Policy in America,* 236.

22 By 1978, inflation in the United States: "1978 Inflation Rate 9% in U.S.," *New York Times,* January 25, 1979.

22 As Vietor has detailed: Vietor, *Energy Policy in America,* 237–39.

22 nation's first energy czar: "Nixon's Decisive New Energy Czar," *Time Magazine,* December 10, 1973.

24 right down to the level of retail distribution: Vietor, *Energy Policy in America,* 236.

25 President Ford proposed a new plan: "Ford Plans Decontrol of Prices on Oil," *Deseret News,* February 17, 1975.

25 These tertiary technologies: "Billions of Barrels of Oil, Waiting," *New York Times,* November 27, 1977.

26 didn't know enough to write a price fixing order: Marshall, *Done in Oil,* 37.

28 In the winter of 1976–77, curtailments cost: Tussing and Tippee, *Natural Gas Industry,* 194.

29 most difficult public policy issues: Committee on Energy and Natural Resources, *Natural Gas Pricing Proposals,* iii.

31 neither Washington nor the industry anticipated: Tussing and Tippee, *Natural Gas Industry,* 194.

40 One such speculative case: Paul Erdman, "The Oil War of 1976—How the Shah Won the World," *New York Magazine,* December 2, 1974.

40 the execution of the Iranian generals: "Secret Police Head and 3 Others in Iran Said to Be Executed," *New York Times,* February 16, 1979, and "20 More Shah Aides Said to Face Death in Start of a Purge," *New York Times,* February 17, 1979.

41 result in over a million deaths: "8-Year Gulf War: Victims but No Victors," *New York Times,* July 25, 1988.

41 leadership by inducing fear: Coughlin, *Saddam,* 197.

41 world markets were deprived: "OPEC Had Its Beginnings Defending $1.80 Oil Price," *Montreal Gazette,* June 29, 1979.

Chapter 3. Reorganization and Consolidation

44 world-class petroleum accumulation: Texas Railroad Commission testimony of Stenzel and other Marathon witnesses, November 7, 1977.

47 total U.S. discoveries of oil: "Outlook Promising for Future Oil Discoveries," *Petroleum Independent Magazine,* February 1980.

49 able to grow their reserves: First Boston Report, "Recent Petroleum Mergers and Acquisitions"; presentations before the Society of Petroleum Engineers of AIME 1983 SPE Hydrocarbon Economics and Evaluation Symposium, March 1983, 6.

50 with the following introduction: Paine, *Oil Property Valuation,* 15.

58 partial tender for 15 percent: "It's Time to Make a Deal," *Texas Monthly,* October 1982, 228.

58 how to get Boone to step aside: Pickens, *Boone,* 65–70.

61 outgrowth of his early success: Lenzner, *The Great Getty,* 35, 53.

62 excesses of capitalism: "Tass Condemns Texaco's Plan to Acquire Getty Oil," *Wall Street Journal,* January 10, 1984.

62 $10 billion judgment: Wasserstein, *Big Deal,* 191, 192.

Chapter 4. Turning Points

68 examples of distortions: Gustafson, *Crisis amid Plenty,* 86, 87, 95, 104, 111, 115, 169; Economides and D'Alea, *From Soviet to Putin,* 317, 318.

73 exceptionally prescient article: William M. Brown and Herman Kahn, "Why OPEC Is Vulnerable," *Fortune,* July 14, 1980, 66–69.

74 King Fahd announced: "Will Oil Prices Hold at $18? A Game the Saudis Can't Afford," *New York Times,* February 8, 1987.

75 members of OPEC were ready: "Crude Prices Hover Below $10 a Barrel on Some Markets," *Los Angeles Times,* July 11, 1986.

76 Alan (who is now deceased) suggested: "Is the Crash in Crude Finally Over?" *Barron's,* August 11, 1986, 1, 6–7, 27–31.

79 the deal was priced as recommended: "Business Day Briefs," *New York Times,* September 25, 1987.

81 Bill Lee was later described: "Lucky Bill Lee," *Forbes,* October 14, 1991.

Chapter 5. Game Changers I

87 In May 1990: F. Gregory Gause III speech.

95 reverse Morris Trust: A reverse Morris Trust is a complex provision in the U.S. tax code whereby if financial and technical requirements are met, the asset held by a corporation can be spun off into a new entity without incurring a taxable gain.

97 On November 29, 1997: Energy Information Administration Chronology of World Oil Markets Events, http://www.eia.gov/forecasts/aeo/assumptions/pdf/0554(2012).pdf.

99 in March 1999 Saudi Arabia agreed: Ibid.

99 "Drowning in Oil": *Economist,* cover story, March 6, 1999.

99 "The Death of Equities": *Business Week,* cover story, August 13, 1979.

100 "Goofs: We Woz Wrong": *Economist,* December 16, 1999.

100 In late October 1998: "Mideast Reopening Portends Massive Change for Industry," *Petroleum Intelligence Weekly,* November 16, 1998; *Wall Street Journal,* November 12, 1998.

104 Daniel Yergin has: Yergin, *The Quest,* 89–105.

112 diminish U.S. influence: *Wall Street Journal,* editorial page, September 11, 2012.

114 Saudis have to deal with the challenges: "Saudi, Chinese Agree to Landmark Energy Accord," *Oil and Gas Journal,* February 6, 2006.

115 China and India announced: "Oil Firms in China and India Pull Closer," *Wall Street Journal,* June 20, 2012.

Chapter 6. Game Changers II

123 well over one-half of the world's producing countries: Bower, *Oil,* 305.

124 This process is described: Sampson, "Libyan Ultimatum," chapter 10 in *The Seven Sisters,* 208–29.

124 The subsequent full playing out: Yergin, "The Hinge Years: Countries versus Companies," chapter 28, and "The Oil Weapon," chapter 29, in *The Prize,* 577–87 and 588–609, respectively.

125 decline rate in the existing global producing base: Yergin, *The Quest,* 239.

126 In the view of some observers: Jim Mulva, now the former CEO of ConocoPhillips, and Christophe de Margerie, CEO of Total.

126 Our ignorance is not so vast: M. King Hubbert website, http://www.hubbertpeak.com.

129 many operators of these new facilities: "Natural Gas Glut Drives Wave of Export Projects," *Wall Street Journal,* October 5, 2012.

130 there may be political considerations: "Natural Gas Exports, Maybe," *Wall Street Journal,* May 21, 2013; "Gas Export Opponents Ignite U.S. Shale Debate," *Financial Times,* March 26, 2013.

131 this deep geologic trend: Presentation by Energy XXI at EnerCom Conference in San Francisco, February 19, 2013.

131 the end of cheap oil: "Think Gas Is Expensive Now?" *National Geographic,* June 2004.

132 outlook for future production from Saudi Arabia: "Rebutting the Critics," *Oil and Gas Journal,* May 17, 2004.

135 what we are experiencing: Senate testimony of Michael Masters, May 20, 2008.

135 According to Bart Chilton: "Oil-Pricing Firm Fought a Push for Regulation," *Wall Street Journal,* May 16, 2013.

138 explicitly criticized by senior executives: "Big Oil Turns Its Back on BP," *Business Week,* July 19–25, 2010.

138 industry-wide systemic problems: "White House Probe Blames BP, Industry in Gulf Blast," *Wall Street Journal,* January 6, 2011.

138 began to take remedial action: "Salazar Looking to Beef Up Regulatory Muscle," *Upstream,* October 7, 2011.

139 Senator Mary Landrieu . . . was particularly outspoken: "Industry Unhappy at US Gulf Uncertainty," *Upstream,* November 26, 2010.

139 a recent analysis by RBN Energy confirms: "Deepwater Gulf of Mexico Rebounding from Macondo," *Oil and Gas Financial Journal,* April 2013, 14.

140 Nine of the largest companies: "Giants Join Forces to Beat Blowout," *Upstream,* May 20, 2011.

140 In addition, there is a second: "Companies Form 100 Strong Well Control Response Team," *Upstream,* March 29, 2013.

141 federal judge Martin Feldman: "Court Tells US to Act on Permits," *Wall Street Journal,* February 18, 2011.

141 This was titled the "Restarting American Offshore Leasing Now Act": "House Passes Bill to Speed Sale of Offshore Oil Leases," *New York Times,* May 6, 2011.

142 require the Secretary of the Interior: "Lamborn Advocates 30 Day Limit to Act on Permits," *Denver Post,* May 13, 2011, 6A.

142 the Gulf of Mexico's remaining undiscovered: U.S. Geologic Province Petroleum Assessment Summaries, http://energy.usgs.gov/OilGas/AssessmentsData/NationalOilGasAssessment/USBasinSummaries.aspx?provcode=5047.

143 Shell Oil's efforts to explore: "Shell Drill Plan Nears EPA Nod," *Wall Street Journal,* May 14, 2011; "Shell Is Likely to Receive Permits for Oil Drilling Off Alaska," *New York Times,* June 27, 2012; "BP, Shell Run into Alaska Offshore Setbacks," *Wall Street Journal,* July 11, 2012; "Shell's Arctic Delays Deter Others," *Financial Times,* September 5, 2012; "For Shell, Wait Til Next Year in Arctic," *Wall Street Journal,* November 1, 2012; "EPA Citation Further Muddles Shell's Arctic Plans," *Wall Street Journal,* January 12, 2013.

144 environmental activists' opposition: "An Alaskan Challenge for 'All of the Above Energy,'" *Wall Street Journal,* February 2 and February 3, 2013.

Chapter 7. Alternative Energy Options

154 fossil fuels remain dominant in the global energy mix: International Energy Agency, *World Energy Outlook 2012,* Executive Summary, 1.

155 (NREL) originally began operating as: NREL website, www.nrel.gov.

157 contemplate the merits of: Sandalow, *Plug-In Electric Vehicles,* 69–75.

158 using today's state-of-the-art technology: "NREL Continues to Grow with Its Newest Facility Opening Later This Year," *Denver Business Journal,* July 13 19, 2012.

161 the winning bid of $2.5 million: "Ethanol Plant Could Face Wrecking Ball," *Wall Street Journal,* May 1, 2013.

161 Lawrence Mone . . . has observed: "How to Avoid Making the Energy Boom Go Bust," *Wall Street Journal,* op-ed page, August 24, 2012.

164 coal generation of electricity: "Shale Gas Boom Fuels Big Drop in U.S. Carbon Emissions," *Financial Times,* May 24, 2012.

164 upheld the EPA's moves to restrict: "Court Backs EPA on Warming," *Wall Street Journal,* June 27, 2012, A1 and A6.

165 As pointed out: "U.S. Regulator Halts Nuclear-Plant Licensing," *Wall Street Journal,* August 7, 2012.

166 U.S. wind production credit is estimated: "The Winds of Washington," *Wall Street Journal,* December 26, 2012.

Chapter 8. Unconventional Fossil Fuels

169 advances were pioneered by George Mitchell: "Cracking Energy's Puzzle Dig Deep: George Mitchell's Fracking Lifted the Industry," *Investor's Business Daily,* November 7, 2011; "Texas Oilman's Idea Leads to Shale Boom," *Oklahoman,* September 30, 2012.

169 dozen other new shale plays: Don Lyle, "Shale Gas Plays Expand," *E&P Magazine,* March 17, 2007.

171 impressive potential for production growth: Massachusetts Institute of Technology, *Future of Natural Gas,* 30–37.

175 there is now reason to believe: Ibid.

177 Pickens has proposed: Pickens Plan website, www. PickensPlan.com. See tab "About the Plan," section entitled "Natural Gas Is a Domestic Fuel That Can Free Us from OPEC Oil."

178 Of the U.S. vehicle population, total energy usage: NGVAmerica website, www.ngvc.org.

180 drilling companies are beginning to convert: "Drillers Shift to Use of Natural Gas," *Wall Street Journal,* December 26, 2012.

181 Another innovation is at an even earlier stage: "Railroad Test Switch to Gas," *Wall Street Journal,* March 6, 2013, B1.

182 MIT study group recommended: Massachusetts Institute of Technology, *Future of Natural Gas,* 133.

185 In subsequent announcements: Continental Resources press release, December 3, 2012.

185 an updated May 2013 assessment: Presentation at the Hart Energy DUG Conference in Denver.

186 One example is the announcement: "First Reserve, Triangle Form Pipeline Venture," *Wall Street Journal,* October 1, 2012, B3.

187 Burlington Resources and other companies: "Horizontal Drilling Grows in Williston," *Oil and Gas Journal,* November 6, 1989.

190 This conclusion has been indicated publicly: Michelle Thompson, "The Eagle Is Not Landing," *Midstream Business Operator,* October 2012, 24.

Chapter 9. Upcoming Issues and Challenges in the U.S. Energy Outlook

194 Mason's focus was on the policies: "Time for a Cease-Fire in the War on Oil," *Wall Street Journal,* op-ed page, April 25, 2011.

196 the surge in North American oil production: "Gushing Forecasts for Oil," *Wall Street Journal,* May 15, 2012.

197 clear potential for a large improvement: International Energy Agency, *World Energy Outlook 2012.*

197 In September 2013, the Wall Street Journal reported that Saudi Oil Minister Ali Naimi: "U.S. Set to Pass Russia in Production of Fuel," *Wall Street Journal,* September 13, 2013.

198 In addition, Tom Donilon: "Obama Backs Rise in U.S. Gas Exports," *Financial Times,* May 6, 2013.

198 As observed by Ed Crooks and Geoff Dyer: "Strength in reserve," *Financial Times,* September 6, 2013.

200 the likelihood remains: "Resource-Rich Canada Looks to China for Growth," *Wall Street Journal,* May 14, 2012.

201 flawed incentives also have been counterproductive: "Drop in Fuel Use Reignites Debate over U.S. Ethanol," *Financial Times,* March 12, 2013.

204 regulatory authority has acknowledged: public statement by the State of Colorado Oil and Gas Conservation Commission to correct several errors in the film *Gasland*'s portrayal of Colorado incidents.

204 there was fabricated injection: "OPEC Finances Anti-fracking Flick," *Investor's Business Daily,* December 24, 2012. According to the Heritage

Foundation, significant funding for *Promised Land* was provided by an Abu Dhabi–based source interested in promoting antifracking sentiment.

204 committee consisted of prominent citizens: The members of the committee were John Deutch, Institute Professor at MIT (chair); Stephen Holditch, Department of Petroleum Engineering at Texas A&M; Fred Krupp, president of the Environmental Defense Fund; Kathleen McGinty, an environmental leader; Susan Tierney, an energy and environmental consultant; Daniel Yergin, chairman of IHS Cambridge Energy Research Associates; Mark Zoback, a professor of Geophysics at Stanford University. U.S. Department of Energy, Shale Gas Production Subcommittee 90-Day Report, August 18, 2011.

209 The global energy map is changing: International Energy Agency, *World Energy Outlook 2012*, 23.

GLOSSARY

"Bail-out zone" A secondary formation completion objective in a drilled well, usually likely to yield only enough petroleum to partially recover the cost of drilling a well.

Black swan event A widely perceived low- or no-probability occurrence that thus has high impact when it does actually happen. This term stems from an old European belief that there was no such thing as a black swan, because none had ever been seen by Europeans until **Australia was discovered.**

Blowout preventer A heavy well casing head control device consisting of rams or gates that can be quickly closed hydraulically around drill pipe to shut off the flow of high-pressure fluids in an emergency.

"Bright spot" geophysical analysis A geophysical technique to determine the likelihood that a seismically defined image is exhibiting attributes that make it likely to contain hydrocarbons.

Butane A gaseous hydrocarbon molecule (C_4H_{10}) under ambient atmospheric conditions. However, it is often blended with gasoline (especially in winter) for easy starting and better acceleration.

Casing pipe Heavy-grade steel pipe that is set at the top of a wellhead to seal fluids off from the hole. It also serves to prevent the hole from caving in. Depending on the nature of the surface formation, multiple strings of casing pipe may be used with one inserted inside another.

Condensate A light liquid hydrocarbon, typically exhibiting an A.P.I. gravity of 45 degrees or higher. The molecules contain carbon ranging from C_5 to C_{10}. It is readily refined into the lighter transportation fuels.

Conventional oil Typically oil that is produced from a vertical well drilled into a porous rock formation exhibiting relatively high permeability. In such cases, the oil has been generated in a different source rock and migrates into the pore spaces of the producing formation.

"Corner shots" Low-risk drilling prospects usually closely offsetting existing producing wells.

Corporate activist A financial operator who has the goal of initiating or precipitating a financial reorganization designed to provide short-term rewards to shareholders.

"Crown jewel" option The right to purchase a key or major corporate asset under certain conditions occurring or not occurring in connection with a merger. This has been declared illegal in current case law.

Cumulative oil production The total volume of production of hydrocarbons over the period from inception to the point of the current calculation.

Deepwater Typically involves drilling activity in water depths between 1,000 and 10,000 feet.

"Dry" natural gas Predominantly methane that is not unassociated with oil or condensate and involves very limited amounts of natural gas liquids (NGLs).

Electrical well logs The result of a process by which instruments are lowered into a drilled formation to yield measurements of the electrical conductivity of a formation as an indicator of its hydrocarbon prospectivity.

Electric vehicles Vehicles powered by electric motors and heavy-duty batteries.

Enhanced recovery oil Oil produced as a result of waterflooding, steamflooding, carbon dioxide injection, or other supplementary recovery techniques.

Ethane A simple hydrocarbon (C_2H_6) typically associated with petroleum. It is a gas at normal atmospheric conditions and is a useful feedstock material for petrochemical processes.

Executive session A board meeting without the presence of members of a company's management.

Export quotas Pre-agreed limits on the amount of oil to be exported for a given period by various members of OPEC.

Finding costs per barrel The amount of capital and exploratory spending to find and bring to production an oil field divided by the estimated reserves of the accumulation.

"Frontier" A term referring to regions or exploratory trends that have been either lightly explored or not even previously evaluated.

Frozen wellhead A condition in which a well has ceased to produce hydrocarbons, caused by the freezing of water associated with the petroleum.

Geomagnetic surveys Typically, these are airborne screenings to determine subsurface magnetic trends that may be indicative of hydrocarbon trapping conditions.

Greenhouse gas emissions Gases released into the atmosphere that tend to reflect heat back to the earth. Among others, these include carbon dioxide and methane.

Greenmail Payments made to a company in return for the promise not to promote a restructuring or corporate takeover.

Hostile bid An unsolicited offer to purchase part or all of the shares of a publicly traded company.

Hybrid electric A vehicle with both a conventional petroleum-fueled engine and an electric motor drive that can readily switch between these means of propulsion, as appropriate.

Impermeable A term used to describe tight rock formations that are not naturally conducive to the flow of hydrocarbons (gases or liquids).

Implied per barrel value of oil and gas reserves behind each share of stock The debt plus equity value of a producer divided by its proved reserves.

Interstate A term referring to commercial activity occurring across more than one state of the United States.

Land grant The historical award of federal acreage to transcontinental railroads as an incentive to build the connecting rail infrastructure. The awards consisted of alternating one-mile-square blocks forming a twenty-mile-wide checkerboard along the route constructed.

Liquefaction The process of converting methane from a gas to a liquid by reducing its temperature to -258° Fahrenheit. This results in a six-hundred-fold reduction in the volume of the methane.

Master limited partnership (MLP) An enterprise established to hold income-producing assets and to distribute that income to the MLP shareholders on a pass-through basis.

Natural gas A hydrocarbon gas mixture consisting mainly of methane molecules (CH_4).

Natural gas liquids (NGLs) A form of processed natural gas that may consist of one or more hydrocarbon compounds including propane, ethane, butane, iso-butane, and pentane.

Neutral Zone concession A disputed territory between Saudi Arabia and Kuwait in which each country shares 50 percent of the production.

Oil shale Also known as kerogen, a less than fully matured oil.

Overthrust Belt A subsurface buckling of the earth's crust that can create structures conducive to trapping oil or natural gas.

Pentane A paraffinic hydrocarbon (containing five carbon atoms) that is a liquid at normal atmospheric conditions.

Propane A flammable gas at normal atmospheric conditions (C_3H_8) that is readily used to heat homes as well as cook and refrigerate food.

Proved reserves Oil that has been discovered and defined as to how much is economically recoverable.

Recovery rates The percentage of oil or gas in a reservoir that is expected to be produced using currently available technology.

Regasification (regas) The conversion of liquefied natural gas (LNG) back to natural gas by rewarming the liquid to ambient temperature.

Reserve replacement rate The amount of hydrocarbons added by an enterprise from new investment in a given time period divided by the amount of hydrocarbons produced in the same period.

Residual oil in place The oil remaining in a reservoir after recovery of reserves by currently available technology.

Seismic technology The interpretation of subsurface structural features or stratigraphic changes via the introduction and received reflection of sound waves propagated into the earth's crust.

Sell side That portion of the financial market participants involved with distributing securities to investors and institutions.

Shale oil Fully thermally matured oil that is producible from an effectively stimulated subsurface formation.

"Smart" electrical grid This provides for the two-way flow of electricity via a computer-optimized distribution and storage of renewable energy sources.

Standardized measure of discounted future revenues The future net revenues of petroleum reserves produced over their expected lifetime discounted at 10 percent to the present value.

"Stranded" gas Natural gas lacking access to a consuming market.

Stratigraphic traps A petroleum reservoir capable of holding oil or gas because of a change in the character of the rock or a discontinuity in the rock formation (that is, not a structural entrapment).

"Stripper oil" Oil produced in small, marginally economic quantities from a well (typically ten barrels per day or less).

Tectonic plate Large portion of the earth's crust that moves over geologic time and can cause earthquakes as well as a restructuring of subsurface features.

Temperature window The range of subsurface temperatures in hydrocarbon-bearing rock consistent with the occurrence of oil or natural gas. A cooler range of temperature is consistent with the occurrence of oil and a warmer one with natural gas.

Tertiary technologies Processes to produce oil and gas by introducing additional substances and/or energy to facilitate recovery of petroleum over and above that which would flow naturally from a well or field.

"Tight" reservoirs Low-permeability rocks containing hydrocarbons.

Top holes The shallow portions of a well in which surface casing is installed in the early phase of drilling a well. See casing pipe.

Truncation A discontinuity in a subsurface formation that can be conducive to trapping hydrocarbons.

Ultradeep A term used to refer to petroleum prospective formations deep in the earth's crust (typically 25,000–30,000-plus feet).

Unconformity A layer of reservoir rock deposited on deformed or eroded rock such that a seal is formed and facilitates the entrapment of petroleum.

U.S. flag tankers Oil transport ships built and registered to operate between U.S. domestic ports.

"White knight" A potential friendly buyer of a company willing to prevail over a hostile offer.

Windfall profits Profits deemed excessive by a government or regulatory authority, usually due to an unexpected sharp increase in commodity prices.

BIBLIOGRAPHY

Aburish, Said K. *The Rise, Corruption and Coming Fall of the House of Saud.* London: Bloomsbury Publishing, 1994.

Al-Husseini, Dr. Sadad Ibrahim. "Rebutting the Critics: Saudi Arabia's Oil Reserves, Production Practices Ensure Its Cornerstone Role in Future Oil Supply." *Oil and Gas Journal,* May 17, 2004.

Allard, Kenneth. *Business as War: Battling for Competitive Advantage.* Hoboken, N.J.: John Wiley and Sons, 2003.

Ambrose, Stephen E. *Nothing Like It in the World: The Men Who Built the Transcontinental Railroad, 1863–1869.* New York: Simon and Schuster, 2001.

Aspen Institute. *Climate Change and the Electricity Sector: Energy Markets and Global Politics: 2006 Forum on Global Energy, Economy and Security.* J. Bennett Johnston, chair, and Leonard L. Coburn, rapporteur. Queenstown, Md.: Aspen Institute, 2006.

———. *Climate Change and the Electricity Sector: 2008 Energy Policy Forum.* Jonathan Lash and Jeff Sterba, co-chairs, and Paul Runic, rapporteur. Queenstown, Md.: Aspen Institute, 2008.

Associated Press. *Lightning out of Israel: The Six-Day War in the Middle East.* Racine, Wis.: Western Printing and Lithographing, 1967.

Athearn, Robert G. *Union Pacific Country.* New York: Rand McNally, 1971.

Bain, David Haward. *Empire Express: Building the First Transcontinental Railroad.* New York: Viking Penguin, 1999.

Baker, James A., III, and Lee H. Hamilton, co-chairs. *The Iraq Study Group Report: The Way Forward—A New Approach.* New York: Vintage Books, 2006.

Berenson, Alex. *The Number: How the Drive for Quarterly Earnings Corrupted Wall Street and Corporate America.* New York: Random House, 2004.

Berndt, Ernst, Edward Erickson, Basil Kalymon, James McKie, Walter Mead, David Quirin, Russell Ulher, Michael Walker, Campbell Watkins, and Herbert Winokur, Jr. *Oil in the Seventies: Essays on Energy Policy.* Vancouver, B.C.: Fraser Institute, 1977.

Bodansky, Yossef. *Bin Laden: The Man Who Declared War on America.* Rockland, Calif.: Prima Publishing, 1999.

Bonner, Bill, and Addison Wiggin. *Empire of Debt: The Rise of an Epic Financial Crisis*. Hoboken, N.J.: John Wiley and Sons, 2005.

Bower, Tom. *Oil: Money, Politics, and Power in the 21st Century*. New York: Grand Central Publishing, 2010.

Brown, Lester R. *Plan B 4.0: Mobilizing to Save Civilization*. New York: W. W. Norton, 2009.

Bryce, Robert. *Gusher of Lies: The Dangerous Delusions of "Energy Independence."* New York: PublicAffairs, 2009.

———. *Power Hungry*. New York: PublicAffairs, 2010.

Clark, James A., and Michel T. Halbouty. *The Last Boom: The Exciting Saga of the Discovery of the Greatest Oil Field in America*. New York: Random House, 1972.

———. *Spindletop*. New York: Random House, 1952.

Committee on Energy and Natural Resources. *Access to Oil: The United States Relationships with Saudi Arabia and Iran*. Washington, D.C.: U.S. Government Printing Office, 1977.

———. *Natural Gas Pricing Proposals: A Comparative Analysis*. Washington, D.C.: U.S. Government Printing Office, 1977.

Committee on Foreign Relations. *Multinational Oil Corporations and U.S. Foreign Policy*. Washington, D.C.: U.S. Government Printing Office, 1975.

Committee on Interior and Insular Affairs. *Geopolitics of Energy*. Washington, D.C.: U.S. Government Printing Office, 1977.

———. *Summary of Responses to Joint Committee Questionnaire on Potential Problems Associated with the Delivery of Crude Oil from Alaska's North Slope*. Washington, D.C.: U.S. Government Printing Office, 1976.

———. *U.S. Energy Resources: A Review as of 1972*. Washington, D.C.: U.S. Government Printing Office, 1974.

Conover, Beth. *How the West Was Warmed: Responding to Climate Change in the Rockies*. Golden, Colo.: Fulcrum Publishing, 2009.

Cooper, Andrew Scott. *The Oil Kings: How the U.S., Iran, and Saudi Arabia Changed the Balance of Power in the Middle East*. New York: Simon and Schuster, 2011.

Copp, E. Anthony, ed., *World Petroleum: The Economic of Current Pricing and Supply Policies*. London: Seminar Sponsored by Salomon Brothers, 1975.

Coughlin, Con. *Saddam: King of Terror*. New York: HarperCollins, 2002.

De Blij, Harm J. *Why Geography Matters: Three Challenges Facing America—Climate Change, the Rise of China, and Global Terrorism*. New York: Oxford University Press, 2005.

Deutch, John M. *The Crisis in Energy Policy*. Harvard, Mass.: Harvard Press, 2011.

Donahue, Jack. *Wildcatter: The Story of Michel T. Halbouty and the Search for Oil*. New York: McGraw-Hill, 1979.

Economides, Michael J., and Donna Marie D'Alea. *From Soviet to Putin and Back: The Dominance of Energy in Today's Russia*. Houston, Tex.: Energy Tribune Publishing, 2008.

Energy Policy Project of the Ford Foundation. *A Time To Choose: America's Energy Future*. Cambridge, Mass.: Ballinger Publishing, 1974.

Foreign Affairs. "Changing Course in the Persian Gulf." Vol. 76, no. 3 (May/June 1997).

———. "The Politics of Peace Paralysis." Vol. 77, no. 4 (July/August 1998).

Foster, Peter. *The Blue-Eyed Sheiks: The Canadian Oil Establishment*. Toronto, Ont.: Collins Publishers, 1979.

Friedman, George. *The Next 100 Years: A Forecast for the 21st Century*. New York: Knopf Doubleday, 2010.

Goodstein, David. *Out of Gas: The End of the Age of Oil*. New York: W. W. Norton, 2004.

Gore, Al. *An Inconvenient Truth: The Planetary Emergency of Global Warming and What We Can Do about It*. New York: Rodale, 2006.

Graves, John. *Fracking: America's Alternative Energy Revolution*. Ventura, Calif.: Safe Harbor International Publishing, 2012.

Gustafson, Thane. *Crisis amid Plenty: The Politics of Soviet Energy under Brezhnev and Gorbachev*. Princeton, N.J.: Princeton University Press, 1991.

Harston, Robert. *America's Energy: A Small Book about a Big Problem*. Denver, Colo.: privately published, 2010.

Hart Energy. *Hydraulic Fracturing: The Techbook*. A supplement to E&P. Houston: Hart Energy Publishing, 2012.

Hirsch, Robert L., Roger Bezdek, and Robert Wendling. *Peaking of World Oil Production: Impacts, Mitigation, and Risk Management*. Washington, D.C.: U.S. Department of Energy, National Energy Technology Laboratory, 2005.

Hofmeister, John. *Why We Hate the Oil Companies: Straight Talk from an Energy Insider*. New York: Palgrave Macmillan, 2010.

Howarth, Robert W., Renee Santoro, and Anthony Ingraffea. *Methane and the Greenhouse-Gas Footprint of Natural Gas from Shale Formation*. Ithaca, N.Y.: School of Civil and Environmental Engineering, Cornell University, 2011.

Huler, Scott. *On the Grid: A Plot of Land, an Average Neighborhood, and the Systems That Make Our World Work*. New York: Rodale Press, 2010.

Institute for Contemporary Studies. *No Time to Confuse: A Critique of the Final Report of the Energy Policy Project of the Ford Foundation, "A Time to Choose America's Energy Future."* San Francisco, Calif.: Institute for Contemporary Studies, 1975.

International Energy Agency. *World Energy Outlook 2009.* Paris, France: OECD/IEA, 2009.

———. *World Energy Outlook 2012.* Paris, France: OECD/IEA, 2012.

James A. Baker III Institute for Public Policy of Rice University and the Council on Foreign Relations. *Strategic Energy Policy: Challenges for the 21st Century.* New York: Council on Foreign Relations, 2001.

Jenks, Philip, and Stephen Eckett. *Rules: The Global-Investor Book of Investing Rules; Invaluable Advice from 150 Master Investors.* Hampshire, U.K.: Harriman House, 2001.

Learsy, Raymond J. *Oil and Finance: The Epic Corruption from 2006 to 2010.* Bloomington, Ind.: privately published, 2011.

Lenzner, Robert. *The Great Getty: The Life and Loves of J. Paul Getty—Richest Man in the World.* New York: Crown Publishers, 1985.

Lohr, Steven. "The Great Oil Rush of the 80s." *New York Times Magazine,* August 30, 1981. New York edition, section 6.

Lovins, Amory B. *Soft Energy Paths: Toward a Durable Peace.* Cambridge, Mass.: Ballinger Publishing, 1977.

Margonelli, Lisa. *Oil on the Brain: Adventures from the Pump to the Pipeline.* New York: Doubleday Broadway Publishing Group, 2007.

Marshall J. Howard, II. *Done in Oil: An Autobiography.* College Station: Texas A&M University Press, 1994.

Massachusetts Institute of Technology. *The Future of Natural Gas: An Interdisciplinary MIT Study.* Ernest J. Moniz, chair, and Henry D. Jacoby and Anthony J. M. Meggs co-chairs. Cambridge, Mass.: MIT Energy Initiative, 2011.

McKittrick, Sterling, Jr., and Rosario S. Ilacqua. *Oil Analysts: A Historical Perspective. Part I: To 1989,* by Sterling McKittrick, Jr., and Rosario S. Ilacqua. *Part II: 1989–1999,* by Rosario S. Ilacqua. Concord, Mass.: National Association of Petroleum Investment Analysts, 1999/2000.

Michaels, Patrick J., and Robert C. Balling, Jr. *Climate of Extremes: Global Warming Science They Don't Want You to Know.* Washington, D.C.: Cato Institute, 2009.

Mostert, Noel. *Supership.* New York: Alfred A. Knopf, 1974.

National Geographic. "The Big Thaw: Ice on the Run, Seas on the Rise." June 2007.

———. "The End of Cheap Oil." June 2004.
———. "Growing Fuel: The Wrong Way, the Right Way." October 2007.
Oren, Michael B. *Six Days of War: June 1967 and the Making of the Modern Middle East*. New York: Oxford University Press, 2002.
Paine, Paul. *Oil Property Valuation*. New York: John Wiley and Sons, 1942.
Paul, Bill. *Future Energy: How the New Oil Industry Will Change People, Politics, and Portfolios*. Hoboken, N.J.: John Wiley and Sons, 2007.
Petzinger, Thomas, Jr. *Oil and Honor: The Texaco-Pennzoil Wars; Inside the $11 Billion Battle for Getty Oil*. New York: G. P. Putnam's Sons, 1987.
Pickens, T. Boone, Jr. *Boone*. Boston, Mass.: Houghton Mifflin, 1987.
———. *The First Billion Is the Hardest*. New York: Crown Publishing Group, 2008.
Porter, Stanley P., with Arthur Young & Company. *A Study of the Subjectivity of Reserve Estimates and Its Relation to Financial Reporting*. S.l.: s.n., 1980.
Rand, Christopher T. *Making Democracy Safe for Oil*. Boston: Atlantic–Little, Brown Books, 1975.
Sampson, Anthony. *The Seven Sisters: The Great Oil Companies and the World They Shaped*. New York: Viking Press, 1975.
Sandalow, David. *Freedom from Oil: How the Next President Can End the United States' Oil Addiction*. New York: McGraw-Hill, 2008.
———. *Plug-In Electric Vehicles: What Role for Washington?* Washington, D.C.: Brookings Institution Press, 2009.
Schweizer, Peter. *Victory: The Reagan Administration's Secret Strategy That Hastened the Collapse of the Soviet Union*. New York: Atlantic Monthly Press, 1994.
Scott, Mary Woods. *Near Hits ??? in North Dakota*. Grand Forks: North Dakota Geological Survey, 1973.
Shannon, James. *Texaco and the $10 Billion Jury*. Englewood Cliffs, N.J.: Prentice-Hall, 1988.
Simmons, Matthew R. *Twilight in the Desert: The Coming Saudi Oil Shock and the World Economy*. Hoboken, N.J.: John Wiley and Sons, 2006.
Smith, Arthur L. *Something from Nothing: Joe B. Foster and the People Who Built Newfield Exploration Company*. Houston: Bright Sky Press, 2011.
Steward, Dan B., and Frank Paniszczyn, eds. *The Barnett Shale Play: Phoenix of the Fort Worth Basin, a History*. Fort Worth: Fort Worth Geological Society and North Texas Geological Society, 2007.
Taleb, Nassim Nicholas. *The Black Swan: The Impact of the Highly Improbable*. New York: Random House Publishing Group, 2007.
Tertzakian, Peter. *A Thousand Barrels a Second: The Coming Oil Break Point and the Challenges Facing an Energy Dependent World*. New York: McGraw-Hill, 2007.

Turley, J. A. *The Simple Truth: BP's Macondo Blowout*. Littleton, Colo.: Brier Patch, 2012.

Tussing, Arlon R., and Bob Tippee. *The Natural Gas Industry: Evolution, Structure, and Economics*. 2nd ed. Tulsa, Okla.: PennWell Publishing, 1995.

Verleger, Philip K., Jr. *Adjusting to Volatile Energy Prices*. Washington, D.C.: Peterson Institute for International Economics, 1994.

Vietor, Richard H. K. *Energy Policy in America since 1945: A Study of Business-Government Relations*. Cambridge: Cambridge University Press, 1987.

Wasserstein, Bruce. *Big Deal: The Battle for Control of America's Leading Corporations*. New York: Grand Central Publishing, 1998.

White, Richard. *Railroaded: The Transcontinentals and the Making of Modern America*. New York: W. W. Norton, 2011.

Williams, Howard R., and Charles J. Meyers. *Oil and Gas Terms: Annotated Manual of Legal, Engineering, and Tax Words and Phrases*. Albany, N.Y.: Matthew Bender and Company, 1964.

Wilson, Peter W., and Douglas F. Graham. *Saudi Arabia: The Coming Storm*. New York: M. E. Sharpe, 1994.

Yergin, Daniel. *The Prize: The Epic Quest for Oil, Money and Power*. New York: Simon and Schuster, 1991.

———. *The Quest: Energy, Security, and the Remaking of the Modern World*. New York: Penguin Press, 2011.

———. *Russia 2010 and What It Means for the World*. New York: Random House, 1993.

INDEX

References to figures and maps appear in italic type.